"UNNATURAL DISASTERS"

Case Studies of Human-Induced Environmental Catastrophes

Angus M. Gunn

GREENWOOD PRESS
Westport, Connecticut • London

Library of Congress Cataloging-in-Publication Data

Gunn, Angus M. (Angus Macleod), 1920–
Unnatural disasters: case studies of human-induced environmental catastrophes / Angus M. Gunn.
p. cm.
Includes bibliographical references and index.
ISBN 0–313–31999–5 (alk. paper)
1. Nature—Effect of human beings on—Case studies. 2. Disasters—Case studies. I. Title.
GF75.G85 2003
304.2'8—dc21 2002044848

British Library Cataloguing in Publication Data is available.

Library of Congress Catalog Card Number: 2002044848
ISBN: 0–313–31999–5

First published in 2003

Greenwood Press, 88 Post Road West, Westport, CT 06881
An imprint of Greenwood Publishing Group, Inc.
www.greenwood.com

Printed in the United States of America

The paper used in this book complies with the Permanent Paper Standard issued by the National Information Standards Organization (Z39.48–1984).

10 9 8 7 6 5 4 3 2 1

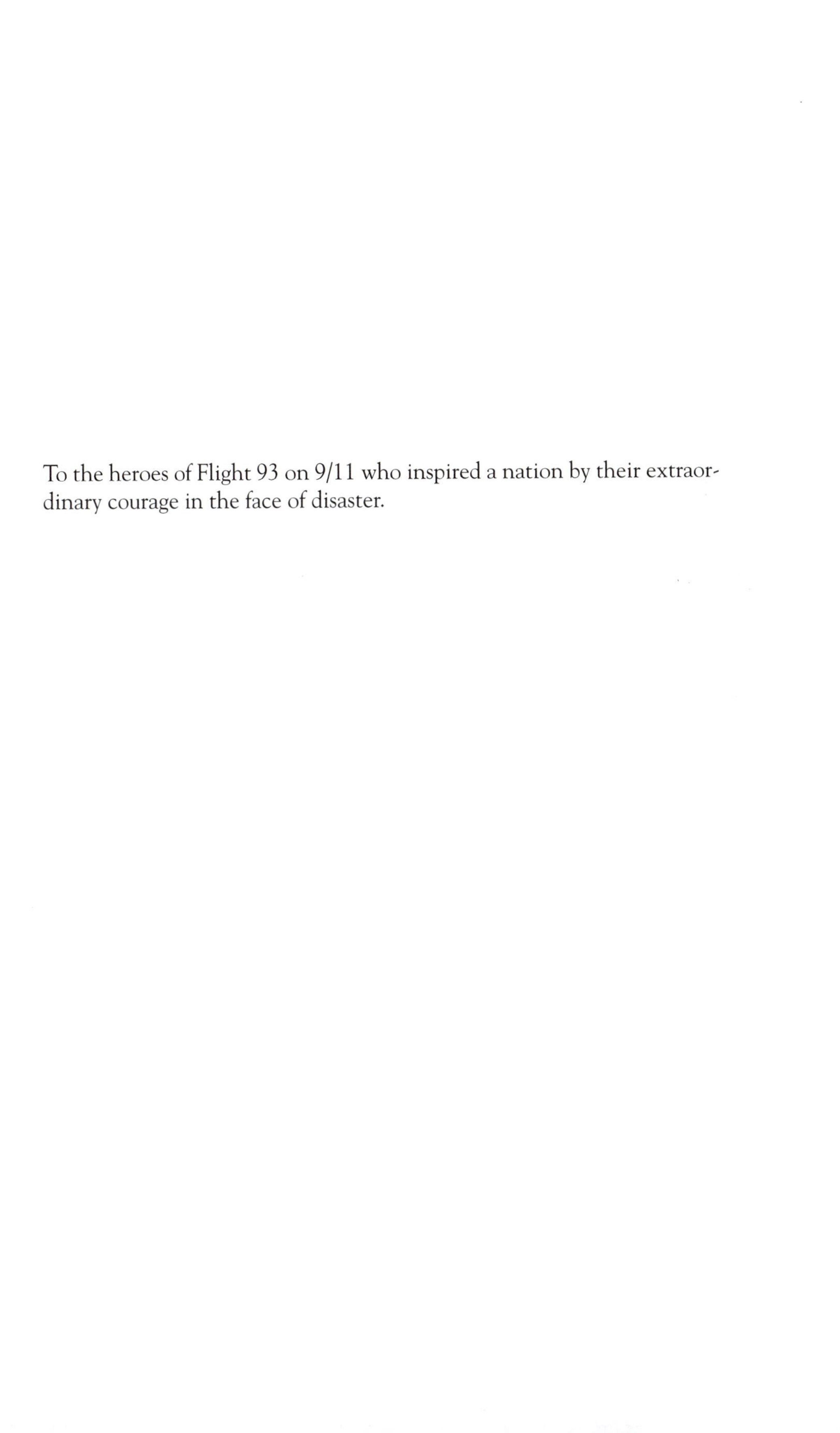

To the heroes of Flight 93 on 9/11 who inspired a nation by their extraordinary courage in the face of disaster.

Contents

Acknowledgments

This book was launched by the creative ideas of Emily M. Birch, Acquisitions Editor of Greenwood Press. I thank her for suggesting that I write about unnatural, human-induced disasters, to parallel the volume I wrote on natural ones, *The Impact of Geology on the United States*. For data on the "Landscape Devastation" case study (Ukraine, 1932), I am most grateful for personal communications with one person who survived the massacres of that time. For their comments and criticisms I am greatly indebted to my family and my colleagues at the University of British Columbia.

Introduction

Awareness of disasters anywhere in the world is vital to the preservation of our global environment. No longer can we be indifferent to events in far-off countries. Many years ago the poet John Donne reminded us of this when he wrote, "No man is an island, entire of itself," and went on to show how all humanity and its environment form one interdependent whole. The tragedy of mercury poisoning in Iraq could have been prevented if those involved had been familiar with Japan's Minamata disease, which had been diagnosed and treated 15 years earlier. In light of this example, it is small wonder that George Santayana wrote, "Whoever forgets the past is doomed to relive it." This book will heighten our awareness of past human-induced—as distinct from natural—disasters so that we can care more responsibly for the environment of planet Earth.

There are natural disasters and there are human-induced ones. The former are familiar—earthquakes, hurricanes, floods, droughts—events over which we believe we have little control and in which we do our best to minimize damage to ourselves and our property. Human-induced disasters appear to be fundamentally different. They are regarded as the results of human error or malicious intent, and whatever happens when they occur leaves us feeling that we can prevent a recurrence. More and more, we find that human activity is affecting our natural environment to such an extent that we often have to reassess the causes of so-called natural disasters. Preventable human error might have contributed to some of the damage.

Take, for example, the great San Francisco earthquake of 1906. The firestorm that swept over the city immediately after the quake, causing far more damage than the direct impact of the earthquake, could have been minimized had alternatives to the city's water mains been in place. The story was similar when the Loma Prieta earthquake struck in 1989. The Marina District of San Francisco, which was known to be unstable and had been severely damaged in 1906, was subsequently developed and built up. When Loma Prieta struck, the district collapsed due to liquefaction. The shaking of the relatively loose soil changed it into a liquid, and buildings sank into it. These secondary effects of human action or inaction are increasingly important considerations in the study of disasters.

ENVIRONMENT

The word *environment* in this book means the natural world around us—the land surface, soils, rocks, water, vegetation, and life forms, together with the atmosphere above. We are very dependent on these things, but our dependence is being greatly reduced, to the point where we ourselves are becoming serious players in the shaping of the natural environment. Global warming is one example of this. Although climatic shifts take place over long periods of time, these shifts are being accelerated by our widespread use of fossil fuels, which raise average temperatures faster than would normally occur.

Our independence from the controlling influence of the natural world is largely related to technology. The tyranny of distance, which used to add costs of travel and communications to everything we did, is enormously reduced by computers, television, and the Internet. University degrees can be earned at home. The most intricate items of clothing or furniture can be manufactured at a distance using computer technology. Our rate of consumption of almost all natural resources has dropped greatly because of the efficiency of new methods and new materials. In some instances we create the new materials we need from atoms. Cars run faster and farther on less fuel, and other machines use less electricity for the same work than they formerly did. As a result we consume less of our irreplaceable supplies of oil and gas.

While the more obvious benefits of technology are easily identified, the dark side must also be acknowledged. Chemical industries worldwide are causing havoc in our air and water, greatly endangering our health. Several case studies in this book illustrate the problem. They are not isolated events. The enormous power and influence that can be wielded by one person raises human behavior to a new level of concern. A single error by one person in a major chemical factory can kill thousands of people. In a nuclear installation such as Chernobyl, one person's mistake can (and did) damage a continent.

We study the environment by the well-known traditional methods of science. In the physical sciences—what we often call the hard sciences, such as physics and chemistry—these tried-and-true traditional methods take the form of carefully observing a few variables while leaving out all the others. From these observations a hypothesis is formed, a tentative theory that might explain what was seen. This theory is then tested for accuracy in further observations of similar variables and modified where needed. As this procedure is repeated, there comes a time when theory and observed data make a good match and we are able to predict future conditions.

Scientific method has become far more complex than this traditional one of hypothesis testing. We live in an age of science and scientific thinking, and the variety of methods of inquiry we employ is as varied as the range of subjects studied. Because the environment includes content from the physical and technological arenas, we can expect to use many modes of scientific inquiry. But what basic principles of scientific inquiry are found in all of these? In essence, clear thinking about what we see or know. In more precise terms, careful, disciplined, logical searching for knowledge about any and all aspects of our environment using the best available evidence.

Both scientific and technological aspects of events feature throughout the book. Acts of terrorism challenged our standards of safety and forced us to initiate new protective systems. Failures in the built environment, such as dams, made us reexamine our theories of materials and geological foundations. In every case study three questions are implicit: What were the human errors and the immediate consequences? What were the long-term effects? What theories or practices had to be reassessed to prevent recurrences?

CASE STUDIES

The human-induced disasters described in this book deal with unique events in which human activities were the principal causes of the tragedies. Sometimes the event was a result of ignorance, and at other times it was due to either error or poor judgment. There are also instances—fortunately, few—in which those responsible deliberately intended to cause destruction. The terrorists who bombed the World Trade Center are examples. The events, scattered throughout the twentieth century and the beginning of the twenty-first century, are drawn from many countries. The United States has the biggest number of a single country because it is an open society and its human-induced accidents are readily identified. Few other countries are as ready to publicize their mistakes, so fewer documented case studies are available from them.

I selected case studies that significantly impacted the environment both immediately and over time. Oil spills are included because they are a continuing threat to the physical environment. Coal-mining deaths were far too numerous in the past, and even today there are too many. The careless handling of poisonous chemical substances, as in Bhopal, are major threats to health. There is no attempt to include all the human-induced disasters of the last hundred years. Rather, selected examples of different kinds of events on land or water or in the air illustrate the various types of disasters.

The unique nature of disasters creates its own rare responses. Emotions are triggered in new ways. There is a sense of isolation felt by both individuals and communities, and this can sometimes lead to passivity or even paralysis if the event is catastrophic. These people are in shock, as if they were recovering from major surgery and feeling that their bodies cannot absorb any more change. Some people react in other ways. There are individual acts of extraordinary bravery. There is also the opposite, looting and assault as people react to what they see as the total breakdown of the social order.

Each case study covers a single event, not a process occurring over time, and so the disasters of wars and epidemics are excluded. The nuclear bombing of Hiroshima, although a war event, is included because of the vast environmental consequences that followed it. Case studies are grouped in chapters by category. Within each chapter, the case studies are alphabetized by geographical location. Each case study opens with a summary overview. Causes, consequences, and cleanup efforts, or remedial actions to ameliorate damage or prevent recurrence, follow. At the end of each chapter, a list of selected readings is provided for further study.

SELECTED READINGS

American Society of Civil Engineers. *Lessons from Dam Incidents*. New York: American Society of Civil Engineers, 1975.

Boorstin, Daniel J. *The Discoverers: A History of Man's Search to Know His World and Himself*. New York: Vintage Books, 1983.

Garrison, Webb. *Disasters That Made History*. New York: Abington Press, 1973.

Kletz, Trevor. *Learning from Accidents*. 3rd ed. Boston: Gulf Professional, 2001.

1

Coal Mining Tragedies

Turtle Mountain Rock Slide
Frank, Alberta, Canada
April 29, 1903

The town of Frank, in Alberta, Canada, dates from 1901, when rich coal seams were discovered in nearby Turtle Mountain and a mine was opened. Its population in the first few years was about 600, and the town was named after the man who opened the mine. Turtle Mountain had been built up in the ancient past with sedimentary rock, mainly limestone, a common geological structure for coal deposits. Layers of coal seams alternated with layers of limestone. In this particular location, mining the coal was especially easy as the seams slanted downward toward the mine entrance so that gravity did most of the work of moving out the coal. A thousand tons of coal was being extracted daily within the first year of operation.

CAUSES

On any given shift, 20 miners extracted coal from a 10-foot-thick seam that, after a year, stretched back 5,000 feet into the mountain. After two years, in which half a million tons of coal had been removed, evidence of the weakening of the sedimentary layers began to appear. Early in 1903, miners noticed on coming to work that some of the supporting pillars were badly splintered, although they had been in place and in good shape at the end of the previous shift. They knew that the slope of the sedimentary layers was steep, so that if a slide were to develop it would soon be accelerated by gravity. However, little was known about local geology, particularly about the degree of cohesion holding adjoining sedimentary layers of rock together, and therefore nothing was done in response to the danger signals.

The immediate trigger that started the disaster was evidently the weather. On the day before the rockslide it was warm and wet, and this was followed by a night with temperatures well below freezing, a common weather pattern in this part of Canada. Water could thus freeze and expand in the small spaces created by inadequate support from the mineshafts and so facilitate rock movement. These are all normal processes of erosion.

CONSEQUENCES

Early in the morning of April 29, 1903, a gigantic slab of limestone rock broke away from Turtle Mountain at the 3,000-foot level. It weighed about 75 million tons, was half a mile wide, and as it crashed down the side of the mountain it broke apart into huge boulders. Most of the town of Frank beneath the mountain was destroyed, and 70 residents were killed. Debris from the slide is still evident today. When so much of the mountainside collapsed, the mine entrance was blocked. Minutes before the collapse, 3 of the 20 men working that shift took loads of coal to the mine entrance, which was 30 feet above the base of the mountain, and sat down there to eat. They were never seen again, buried forever under tons of rock. The remaining 17 miners were trapped inside the mountain 300 feet from the mine entrance.

The Oldman River, which ran between the mountain and the town of Frank, was quickly blocked by the mass of rock, and a lake began to form on it upstream. Water soon rose to the height of the mine entrance and began to seep into the mine. The men inside observed this development, and they also noticed that their air vents—the openings to the outside that provided fresh air—had been closed by the landslide. Aware that their oil lamps would fade out before long, they examined their options. They knew they had to act quickly. Having concluded that removing the mass of material at the mine entrance was impossible, they decided to try cutting a tunnel upward and outward in the hope that they were reasonably close to the surface. Over a period of 12 hours they worked steadily in shifts and finally emerged on the face of the mountain to stare at the destruction below.

The scene before their eyes was terrifying. Where their homes had been there was now a mass of white limestone. All but a small part of the town's center was gone. The falling rock had swept down the mountainside and across a mile of the ground surface, burying everything before it in hundreds of feet of rock. A mile-long section of double-track railway was swept away, as were the main highway and the coal-mining plant. A freight train entering Frank at the moment of the slide was lucky enough to escape. It arrived as the limestone rockslide began and was able to speed past the town before everything crashed around it. All but one of the horses used inside the mine to haul the coal died. A month after the disaster, when the mine entrance was opened, it was found, still alive. Despite efforts to restore it to health, it died a few days later. The total number of human deaths was 76, including the 3 workers who were killed at the mine entrance.

CLEANUP

Subsequent investigations of the tragedy found that the mining company had left inadequate support between mineshafts, thus reducing pressure on higher strata and endangering their stability. As coal mining resumed at Turtle Mountain later in 1903, many of the safety measures that should have been there in 1901 were firmly in place. The Geological Survey of Canada carried out extensive studies in the months following the tragedy and established procedural rules for building supporting pillars within the mine and for reporting signs of earth movements. All went well for about eight years, until ongoing surveys by the Geological Survey revealed a developing fault zone that in time could trigger a massive slide. The mine was closed down permanently. Today there are no buildings where the town center of Frank once stood, but nearby there is an educational and tourist center where people can visit or stay. Displays and lectures on the history of the slide are provided. Footpaths around this center provide close-up views of the rocks that came down the mountain in 1903 and still lie in the places where they originally came to rest.

Aberfan Rock-Mud Slide
Wales, United Kingdom
October 21, 1966

Every coal mine has a waste heap, that is, a deposit of the material brought to the surface with the coal and then left behind as the coal is

Rescue workers continue their search for victims of the Aberfan disaster, in South Wales, where 190 people lost their lives after part of the village school was engulfed by a giant coal avalanche. (© Hulton/Archive)

taken away. Aberfan was located in one of Great Britain's biggest coal-mining areas, South Wales, and waste heaps, or tips, dotted the landscape around it. It was a small village on the banks of the River Taff, and above it towered the steep slopes of Merthyr Mountain, the site chosen for disposal of the wastes from Merthyr Vale Colliery. Perhaps the very familiarity of these heaps made people less sensitive to their danger. The one that overshadowed Aberfan had been there for about eight years. Others dated back to the early 1900s, when mechanical methods of mining became common. It was easy to extract smaller, dirtier coal seams by these methods and as a result there were larger-than-usual heaps of waste. In other countries waste material was often taken underground and used to fill empty spaces in discarded shafts. In Britain the tendency, at least in this part of the country, was to create artificial hills. The hills were later seeded with grass or shrubs to mask their unsightly appearance.

There were no clear regulations in place at the time of the Aberfan disaster for periodical inspection of these waste deposits. Tip number seven, the one that stood above Aberfan, lay across some streams flowing from the mountain, unlike other tips that were placed between streams.

This should have alerted authorities immediately, especially as there had been two small landslides from this tip in 1963. In 1927 a mining expert gave a lecture in Aberfan on the dangers of allowing water to flow beneath tips. He pointed out that if the South Wales mining authorities did not pay for drainage to draw water away from the tips to avoid this danger, they would have to pay for the higher costs of landslides at a later time. The priority among mine managers, however, was coal and little thought was given to the design of tips. The men who selected the location of tip seven had no expertise for making that choice. Waste went up the sides of Merthyr Mountain on rail tracks, then was emptied sideways and backward onto the tip, which gradually rose in height. It was an efficient and cheap storage method. Land costs were negligible, and because of the slope more material could accumulate than on level ground before the heap became so high that it raised questions of instability. By the time of the accident, tip seven was more than 100 feet high.

CAUSES

Unlike previous heaps, the waste deposited on tip number seven in 1965 and 1966 included tailings, the throwaway material from a froth-flotation process. This new method enabled mine owners to extract additional fuel from poor-quality coal that would previously have been discarded as useless. These tailings were a mixture of rock, soil, and chemical residue and they tended to harden when dry. Workers frequently had to soak them before taking them up the slope for ease of handling and dumping. By October 1966 there were huge quantities of tailings on tip seven. The critical factor with this kind of waste is its ability to lower the angle of rest that it finally assumes. Because the tailings were made almost fluid in order to transfer them to the tip they created gentle slopes, usually as low as 4 degrees instead of the 27-degree slope that occurs with dry material. All kinds of regulations had been circulated regarding the danger of gentle slopes giving way. Nevertheless, the low priority given to the care of waste heaps delayed action on this recommendation, and tailings continued to accumulate on tip seven.

In the months before it finally collapsed, tip seven displayed further evidence of its basic instability. Its front edge moved downhill 30 feet in six months, and its interior collapsed several times over the same period of time. These serious indicators of danger were reported but, as in earlier warnings, nothing was done about them. On the morning of the massive slide, one more report of this kind was made by workers on the site. This time a manager decided he would start another tip on the following Monday, three days later. He did not have the opportunity to do so.

CONSEQUENCES

On the morning of October 21, as school began, tip number seven collapsed and swept downward to overwhelm both the school and a row of small houses. Of the 144 people killed, 116 were schoolchildren. The men working near the top of the tip did not see the 140,000 tons of saturated black rock and mud give way because of a heavy fog that day. They also were largely unconcerned because they knew that this tip had a history of giving way from time to time, and nothing very serious had happened when it slipped in 1963. This time, two water mains were broken by the slide, creating a slippery surface that added momentum to the mass movement of material. The slide began shortly after nine in the morning. Intense activity was triggered at once throughout the area, especially once people knew that children were the principal victims. Neighboring coal mines closed down, and miners joined in efforts to free those buried beneath the rubble.

There was no communication with the outside because telephone lines were down. One man whose daughter was reported dead ran to the school from his place of work three miles away and dug all day until forced to stop by darkness. At times there were as many as 5,000 people digging down to reach the school beneath. The high noise level made it difficult to hear cries of help from below. Army units were brought in to control movements of people and cars and to ensure easy passage for ambulances. Rescuers went on working through the night after floodlighting was brought in. All the time there was the fear of further slippage from above because the rain had been heavy for the previous few days.

CLEANUP

For almost three months the official government inquiry into the Aberfan disaster held meetings and heard from numerous witnesses. The National Coal Board (NCB) bore the brunt of the blame, mainly for its continuing indifference to the danger signals that had been obvious for so long to so many. "Their inspectors," said the final report, "were like moles being asked about the habits of birds." No inspector had visited Aberfan for four years prior to the disaster. It was no surprise for the public to be told in the report that the accident was due to "Bungling ineptitude by men who were charged with tasks for which they were totally unfitted." A completely new organizational structure was recommended for the NCB to ensure proper care and safety at all levels. This included the digging of tunnels inside the waste heaps in order to remove the danger of waterlogged slides. The report also proposed the recontouring of these same heaps to remove what had become an eyesore to the villagers.

Another kind of damage became evident in the aftermath of the 1966 landslide: the psychological effects on the villagers. More than 150 residents, half of them adults, needed psychiatric help. They experienced nightmares for years. It was little consolation to them that the NCB was finally doing what it should have done decades earlier. Ten years after the tragedy, Merthyr Mountain looked very different. All the tips had disappeared, either by removal if they were unsightly and dangerous or by recontouring and seeding them with grass and shrubs. A special hillside cemetery and memorial was built at Aberfan commemorating the victims of the landslide. In addition to the financial compensation handed out by the NCB, there were substantial private contributions because of the widespread awareness of the event. Much of this money was used to build a new community center with a concert hall, a swimming pool, and games and hobbies rooms.

Buffalo Creek Collapse
West Virginia
February 26, 1972

In West Virginia, near the border with Kentucky and Virginia, a relatively new method of extracting coal has been used for some time. It is known as strip mining, a process for reaching coal from the surface instead of from underground. It can be done where coal seams are close to the surface. The earth and overlying rock is removed, and the coal seam is then broken up into smaller pieces by explosives and taken to nearby preparation plants for refining. The coarser waste rock is piled up next to the mined area and the finer coal wastes, including the tailings from the preparation plant, are discharged as a thick slurry into an impoundment pond behind the heap of waste rock. Environmentalists are opposed to this method of mining because it destroys large tracts of the land. Mine companies contend that they will restore the ground surface to its original appearance once the mine is closed down.

Mountaintop mining is a variant of strip mining and is common in West Virginia, where the valleys and their streams adjacent to mining operations serve as repositories for the waste materials dumped there. Buffalo Creek, in Logan County, is in one valley used in this way. The sides

Houses jammed together over a railroad track in Buffalo Creek valley shortly after a flash flood raced down the West Virginia stream February 26, 1972, destroying many homes and killing 125 people. (AP/Wide World Photos)

of the valley are steep, and the creek flows through 16 small communities on its 17-mile journey to the town of Man. These communities sprang up over the years in response to the changes occurring in the local coal-mining industry. From the time of the earliest operations in the 1940s there were many changes in the productivity of the mines, with the highest outputs coming in 1970.

CAUSES

The Buffalo Mining Company, a subsidiary of Pittston Coal Company, one of the nation's biggest producers, began using Buffalo Creek as a dump in 1957. Over the succeeding years it made use of additional sites farther upstream. By the early 1970s the creek consisted of a series of dams behind each of which was a pool of black waste slurry. The Buffalo Mining Company, like others in the region, had a history of ignoring environmental safeguards. In 1967 the U.S. Department of the Interior warned West Virginia officials that Buffalo Creek dams and 29 others throughout the state were unstable and dangerous. This warning was the result of the department's intensive studies in the state, inspired by the news of the accident in Aberfan, Wales, a year earlier.

Despite the warning nothing was done to make the dams safer. A drainage bypass system that would protect the residential areas was recommended but not built. By February 1972 concern was widespread as heavy rains deluged the area and the streams and waste pools behind the dams on Buffalo Creek began to rise. These pools by this time had millions of tons of sludgy material on their bottoms, and a half-million gallons of waste liquid kept pouring in daily from the preparation plants. The state of West Virginia had cited the coal company in 1971 when a failure occurred in one dam, but no action was taken to provide an overflow channel.

CONSEQUENCES

On February 26 one big dam on the upper reaches of Buffalo Creek gave way, taking all the others with it. One hundred and thirty million gallons of water and 35 million cubic feet of waste materials roared down the valley at 30 miles per hour. A 25-foot-high tidal wave of slurry, rock, and soil descended on the communities below. There was no time to warn either residents or local authorities. The coal company did not even try. The dam had collapsed at the same spot where it had broken a year earlier. A canyon 45 feet deep in places was carved out of the valley as home after home was washed away. In all, 1,000 dwellings were lost, leaving 6,000 people homeless and 125 dead. Some were able to scramble up the valley sides before the tidal wave reached their location. Others were carried to safety on the tops of their homes as the force of the black water swept the buildings off their foundations. All forms of communication were cut off. This included telephone lines and railroad tracks. The high school in Man was set up as a relief center.

CLEANUP

Some residents were able to come back and build new homes, aided by modest financial assistance from the coal company. Many of the victims, however, were unhappy with the compensation provided and, aided by a lawyer from a major Washington, D.C., law firm, launched a legal claim in federal court. In the course of the legal proceedings it was discovered that this was not the first lethal accident by Pittston and other coal companies in this part of the country. In 1924 a waste pile had given way in Crane Creek, West Virginia, killing seven people and devastating a wide tract of land. Another failure had occurred in neighboring Virginia in 1955 when a Pittston waste pile gave way, destroying homes and property. The end result of the legal challenge was an award of $13,000 to each of the 600 claimants, an award far in excess of the amounts given to those who accepted the coal company's offer.

It took years to rebuild the communities along Buffalo Creek at a cost to the state of $100 million. A memorial monument was built and the 1972 tragedy is remembered annually in a special service. In response to the Buffalo Creek and other disasters, Congress enacted the National Dam Inspection Act, which authorized the United States Army Corps of Engineers to inventory and inspect all nonfederal dams. In addition, President Jimmy Carter issued a memorandum on April 23, 1977, directing a review of federal dam safety activities by an ad hoc panel of recognized experts.

SELECTED READINGS

Anderson, Frank. *The Frank Slide Story*. Calgary, Canada: Frontier Publishing, 1968.

Austin, Tony. *Aberfan: The Story of a Disaster*. London: Hutchinson, 1967.

Miller, Joan. *Aberfan: A Disaster and Its Aftermath*. London: Constable, 1974.

Stern, Gerald M. *The Buffalo Creek Disaster*. New York: Random House, 1976.

2

Dam Failures

Santa Clara River
Ventura County, California
March 13, 1928

In 1913 William Mulholland, Water Bureau superintendent and chief engineer for the City of Los Angeles, designed and supervised the construction of a 235-mile-long aqueduct to bring water from the eastern side of the Sierra Nevada to a point on the Santa Clara River about 30 miles north of downtown Los Angeles. The additional water was harnessed to generate hydroelectricity as it passed through the steep San Francisquito Canyon. Two power stations were built, one at each end of the canyon.

In 1926 Mulholland secured permission to build a dam between the two power stations in order to store a reserve supply of water. Los Angeles had experienced a drought in the early 1920s, and this led to new demands for water reserves. The St. Francis Dam, which was completed in May 1926, was a curved concrete gravity structure approximately 200 feet high. It could store 38,000 acre-feet of water, equal to 12 billion gallons, and was the second largest of the chain of storage basins in the Los Angeles aqueduct system. It would provide enough water in an emergency to meet the city's needs for two months.

Following the devastating San Francisco earthquake of 1906, intensive studies of faults had been conducted all across California, including the

This aerial view shows the broken water barrier of San Francisquito Canyon, CA, after the St. Francis Dam burst on March 13, 1928. The dam's reservoir of 12.5 billion gallons of water poured down the narrow canyon, carrying nearly 500 inhabitants to their deaths. (AP/Wide World Photos)

area around the St. Francis Dam. The San Francisquito Canyon was identified as a fault line connected with the San Andreas, the fault that caused the San Francisco earthquake. Unlike the San Andreas, it was considered to be a dead fault, that is, one that showed no movement over time. It had been described in published reports of the Seismological Society of America years before Mulholland's decision to build a dam there.

A fault line's coinciding with a river is common. It is the weaknesses associated with a fault that create the river in the first place, allowing water to enter and open up the site. Furthermore, the existence of a fault line does not prevent a dam from being built there, and many dams would never have been built in California if this were an absolute deterrent. None of the geologists who studied the proposed site of the St. Francis Dam said that the presence of the fault should prevent a dam. Nevertheless, most of them were concerned about the geological history of this particular site and the strength of the bedrock. They urged the authorities to conduct additional studies to test the suitability of the site, but their advice was rejected.

One representative of the American Society of Civil Engineers who visited the St. Francis Dam site several times before and during construction described some of the problems he found in a report published before March 1928. He noted that as soon as the surface earth cover was removed from the dam site, a clear line of demarcation became visible between the harder rock below and the softer conglomerates above. Between them was a mass of loose material, the end result of ancient slides along the fault line, and he felt that this should be closely examined before construction began.

Harry R. Johnson, consulting geologist for the City of Los Angeles, gave a more detailed site report after the dam's failure. Speaking at a conference of the American Association of Petroleum Geologists, he depicted the character of the bedrock on which the dam's foundation had rested by describing the underside of a huge block of concrete that had been washed some distance from the dam site. All three types of rock were embedded in the concrete in ways that suggested considerable movement had occurred under pressure from the water above.

Evidence mounted that the rock base on which the dam was built was not as thick or dependable as had been assumed. No detailed stress measurements had been made on the foundation materials, and the designers were not aware of the site's geological history. The final failure was clearly due to some kind of collapse brought about by water pressure. Belatedly it was discovered that the fine clay material in the fault zone once wet became a slippery mud. For some days before the failure local observers had seen muddy water coming through cracks in the retaining wall. On the morning of the tragedy a motorist observed large quantities of water escaping. Mulholland was alerted immediately, but after visiting the site he pronounced it safe.

CAUSES

The causes of the disaster, while hinted at in the observations of geologists after the event, was not fully understood for some time. This was partly due to the lower status of the city's consulting geologist when compared with construction engineers. The field of engineering geology did not exist as a specialist discipline in 1928, so for some time there was no suitable person available to conduct rigorous investigations on the site. Many years later extensive studies were conducted by J. David Rogers, an engineering geologist. He was a member of the new specialist field that emerged in the wake of the St. Francis tragedy.

Rogers found that several mistakes had been made in the course of construction, the worst being ignorance of the geological history of the site.

There was a history of massive landslides along the fault line that ran beneath the dam's foundation, some extending back 500,000 years. On the night of the tragedy one of these slides recurred, probably triggered by water pressure. A million cubic yards of material gave way, undermining the east side of the dam and lowering the height of the ground above by several feet. The dam tilted and water began to carve out bigger and bigger escape routes until all support collapsed.

The dam was originally designed for a height of 185 feet, but during construction another 20 feet was added to get more storage but without additional strengthening of the foundation. In the weeks before the collapse there was a period of steady rainfall, so the dam was completely filled on the night of March 13. Other errors at the construction stage were the failure to clean soil from gravel before adding it to concrete and the use of an inferior type of concrete.

CONSEQUENCES

About midnight on March 13, 1928, the dam collapsed with a loud roar. Twelve billion gallons of water in a wave 80 feet high cascaded down into the riverbed of the Santa Clara River, destroying everything in its path as it swept toward the Pacific Ocean, more than 50 miles away. A swath of 79 million acres of land was destroyed and 500 people in the path of the flood were killed, many of them still asleep in their homes when the water hit. When the main torrent of water subsided, much of Ventura County was covered with 70 feet of mud and debris. Damage estimates topped $20 million, a very high figure for that time.

The hydroelectric plant downstream exploded under the impact of water as if it had been bombed. Some residents heard the sound of the advancing flood, but few were able to get out of its way. The natural slope of the land was toward Ventura County and the Pacific Ocean and, fortunately for Los Angeles, a low mountain ridge prevented water from traveling southward. More than a mile below the dam a ranch house was shattered and two occupants were able to hold on to part of the roof as they were swept downstream. Several miles farther on they found themselves sidelined as their temporary raft got caught in some bushes. They held on there until morning.

CLEANUP

Rescuers arrived with supplies early the next morning to tackle the devastated scene. So much debris had been carried down the Santa Clara

River that its bed in several places was thirty feet above its previous level. Morgues were set up wherever they could be while International Red Cross workers provided immediate help for survivors. Precautions were taken to prevent any outbreak of typhus. There was no electricity because the hydroelectric power station below the dam was gone.

Action was taken at once to establish a dam safety agency, the first of its kind anywhere. This new organization required full geological assessments of dam sites before the design stage, including the provision of uniform engineering criteria for testing compacted earth. The tests that were established then are still in use today throughout the world. All dams and reservoirs in the Los Angeles aqueduct system were reassessed in the light of the new regulations, and one outcome of this procedure was an extensive retrofit of one dam.

Teton Dam
Teton River, Idaho
June 5, 1976

Discussions about building a dam on the Teton River began in the early years of the twentieth century and intensified in the 1930s and again in the 1960s. The project envisioned additional water resources for 100,000 acres of land in the Fremont-Madison Irrigation District, local and downstream flood-control benefits, 16,000 kilowatts of electricity, and major recreational facilities. Groundwater pumping in dry years would supplement the water supply when surface flows were inadequate. By the early 1960s, following testing in a number of possible sites, a design was in place.

Teton had been selected in spite of the knowledge that significant seepage would occur. Engineers concluded that the costs of loss from seepage could be contained within the overall budget. There were concerns regarding the rock foundation, but engineers concluded that it was adequate to support the dam and reservoir. The dam would be a 305-foot-high earth-filled structure with a crest length of 3,100 feet. Its total capacity would be at least 200,000 acre-feet of water.

This plan for the Teton basin project was finally approved by Congress in 1964, but for many years thereafter objections from different groups delayed the start of construction. Environmentalists argued that the dam would destroy 17 miles of the Teton River, a popular location for trout fishing, and remove 2,700 acres of deer and elk habitats. The response to these concerns by the U.S. Bureau of Reclamation, the government agency responsible for the project, was that benefits from flood control and irrigation would more than compensate for these losses. The objections from geologists were quite different.

CAUSES

Geologists insisted that the rock on one side of the proposed dam was weaker than on the other side and therefore when compacted by the weight of water it would rupture. Their concerns were not taken very seriously. Even with memories of the 1928 St. Francis Dam failure still alive within the engineering community, the critical roles of geologists were not given adequate weight. A U.S. Geological Survey (USGS) team sent a memo to the Bureau of Reclamation in January 1973, after construction of the dam had begun, expressing concern for the safety of the project. One member of that team, Harold J. Prostka, returned to the site after the tragedy and pointed out in detail why the site should never have been selected. In his view it was geologically young and unstable, with numerous fractures and faults.

Robert Curry, a professor of geology at the University of Montana, argued that poor site selection and an inadequate approach to design and construction led to the failure. He noted that the Bureau of Reclamation's 1961 regional study dealt with site hazards in a broad manner, barely mentioning permeability. Curry was quite sure that the data on which Congress authorized the project was inadequate.

Marshall Corbett, a geologist from Idaho State University, agreed with Curry that the site selection for Teton Dam was wrong. He pointed out that good site selection was important but the good dam sites had long been used up. Steven S. Oriel, another USGS geologist, voiced concern about the inadequacy of scientific information about the site of the dam. In a final report to the Department of Interior, these fears of geologists were confirmed. The report concluded that the design of the dam failed to take adequate account of the foundation conditions and the characteristics of the soil.

These technical findings identified a number of contributing factors but no single factor for the failure of Teton Dam. High on the list of contributing factors were the multiple joints within the volcanic rock beneath and beside the foundation. The silt, calcite, and rubble filling these joints had not been adequately tested for their resistance to water pressure.

The dam, an earthen structure, was built with rock from the bed of the reservoir area. It was well known that this rock had numerous joints, many with fissures as wide as five inches in which silt and other materials had lodged. This was not a matter of concern for the dam structure itself because of the procedures followed in construction to ensure that it was waterproof. It was a different matter in relation to the rock walls bordering the dam. Barely had construction been completed when questions of stability surfaced. Springs were found 600 feet downstream from the dam, the outcome of seepage into groundwater from upstream sections of the Teton River.

Following the years of delay, construction began in February 1972 and was completed by June 1976. The work was speeded up because of all the delays. The reservoir filling rate changed from one foot per day to two feet per day. Filling was largely completed during the months of October and November 1975. Generator installation followed a month later and then the spillway, the arrangement for releasing surplus water, was installed three months after that. On June 1, 1976, Teton Reservoir was raised to its full capacity, containing 10 million cubic yards of water. At the front of the dam it measured 3,100 feet, and it stretched back upstream for 17 miles.

CONSEQUENCES

Early in the morning of June 5, 1976, with the reservoir at its maximum level, a hole was seen to be leaking water near the right abutment on the northwest side of the dam. At 8:00 A.M. water was escaping at a rate of 30 cubic feet per second and by 9:00 A.M. at 50 cubic feet per second. Still, no one felt the problem was serious. This sort of thing often occurred in a new dam. A couple of bulldozers were brought in and loads of loose rocks were pressed into the area of the leak. It soon became clear that there was a huge eroded area beneath the leak. Both bulldozers sank down into it, and their operators were just able to escape before their machines disappeared into the ground. By the middle of the day no one was in any doubt about the danger. Warning messages were sent out to all places downstream. At noon the dam collapsed.

A wall of 80 billion gallons of water rushed down the canyon of the Teton River, taking with it power and telephone lines, power station and pumping plant, and everything else in its path. Millions of cubic yards of mud and rocks were carried away in the flood, and these added greatly to the power of the water when the flow encountered an obstruction. Residents downstream acted as quickly as they could and were able to evacuate the towns of Sugar City, Teton, and Newdale in half an hour. The water rushed through the canyon, largely bypassing Teton, Saint Anthony, and Newdale because they were on high ground.

Outside of the canyon the water spread to a width of about eight miles and sped along at 10 to 15 miles per hour. The rushing water hit the town of Wilford and obliterated it, literally wiping it from the earth. Sugar City, between the two forks of the Teton River, received the full force of a 15-foot-high wall of water crashing down on it. Rexburg was the largest city in the immediate flood area, most of it on the valley floor. The debris-laden water swept past a log mill on the outskirts of town, adding large logs to the flotsam. The logs acted as battering rams and, along with the rushing water, severely damaged buildings throughout the city.

On the evening of June 5, officials of the Mormon Church that owned Ricks College in Rexburg, unaffected because of its location on higher ground, offered to help. Food and housing were supplied to anyone affected by the flood, and the college became a temporary home for many flood victims. On June 6 President Gerald Ford declared Bingham, Bonneville, Fremont, Madison, and Jefferson Counties federal disaster areas, clearing the way for federal assistance.

The water from the Teton Reservoir threatened the venerable American Falls Dam, which lay downstream on the Snake River. In an effort to save it, the outlets on the dam were opened to full bore in the effort to accommodate the incoming flood of water. More water had to be released from the dam than had ever previously been attempted. Authorities were anxious to protect smaller, more vulnerable downstream dams. Floodwaters reached the American Falls Dam on June 7, and fortunately it was able to hold the full volume of water. When the floodwaters receded, the extent of damage began to be assessed. Fourteen deaths were attributed to the tragedy and costs were estimated at $1 billion. After the flood, damage repair became the first priority.

CLEANUP

Cleanup took a full summer. Thousands of volunteers were bused in to help. They boarded the buses early in the morning, worked all day, and rode home at night. Everywhere, they had to cope with the extensive ruination of 100,000 acres of farmland: 13,000 livestock dead, millions of dollars' worth of farm equipment destroyed, 250 stores and offices and 1,000 homes either obliterated or severely damaged. The Idaho National Guard and state police secured the area. The Department of Housing and

Urban Development (HUD) was one of the first to provide relief to the homeless when it brought hundreds of mobile homes to Rexburg. Most impressive of all was the continuing voluntary help from Ricks College. Through the months of June, July, and August it provided accommodation for 2,000 people and handed out 13,000 meals every day. By August 6 all the emergency repairs were completed and the remaining tasks handed over to local authorities.

All that remained of the dam was the central, pyramid-shaped piece. This is still visible today. The dam was never rebuilt. On the 25th anniversary of its failure, the regional director of the Bureau of Reclamation spoke of the developments of the intervening years. Two investigation groups were formed—a Department of Interior team, and an independent panel of respected engineers, geologists, and dam-safety professionals. Both efforts were supported wholeheartedly by the Bureau of Reclamation. Scores of eyewitnesses, contractors, and bureau design and construction personnel were interviewed. On-site forensic studies were conducted, and all correspondence and records extensively reviewed.

The Bureau of Reclamation today has in place several programs for the overall assurance of safe structures. Reclamation dams are inspected annually and in more detail every three years, with a comprehensive evaluation of dam performance under various loading conditions every six years. Instrumentation has been installed at dams to monitor their functioning in ways that visual inspections cannot. Some 13 dams in the Pacific Northwest Region have been modified in this program.

These lessons from the Teton Dam failure continue to be used by Reclamation engineers to train its examiners, designers, construction specialists, and dam operators. When the area around Tacoma, Washington, was hit with a 6.8 magnitude earthquake on February 28, 2001, it immediately triggered on-site visual inspections of 32 Reclamation sites, all within a radius of 316 miles from the earthquake's epicenter. There were no reports of damage to dams and no hydroelectric power operations were affected. Further inspections will be carried out on these dams in the spring of the succeeding two or three years, as they experience differential volumes of water, to ensure that the earthquake did not weaken them.

Lessons were learned and changes were made. These included procedural details such as implementing independent peer review of studies for dams, design changes such as ensuring redundant measures to control seepage and prevent piping, special treatment for fractured-rock foundations, and frequent site visits during construction by the design engineers. Largely because of the Teton disaster, the federal government took action

The deep and narrow Vaiont River valley in the Dolomite Mountains. (© National Oceanic and Atmospheric Administration/Department of Commerce)

to require each federal agency to review its dam-safety activities and to strengthen its dam-safety programs, and Congress passed several acts authorizing a national dam-safety program.

Vaiont River Valley
Italy
October 9, 1963

The Vaiont River valley is located in the Dolomite Mountains of the Italian Alps, about 60 miles northwest of Venice, close to the Italian-Swiss border. A dam had been proposed for this valley in the 1920s to provide hydroelectricity for several rapidly growing northern cities such as Milan and Turin, but it was 1956 before action was finally taken to go ahead with the project. The design called for one of the highest dams in the world, rising to 900 feet above the valley floor. It would stretch across the valley of the Vaiont for 500 feet and impound 150 million cubic yards of water.

CAUSES

There were challenges facing the dam builders from the beginning. The Vaiont River valley is deep and narrow, the reason for the dam's great height. The mountains on either side of the valley, ranging in height from 7,000 to 8,000 feet, are also steep, and engineers wondered about landslides. Massive slides had occurred in the past in the spot where the dam was to be built. Furthermore, the sedimentary rock that formed the Dolomites is composed of layers of limestone, a rock that can be dissolved by water. It was not known how the layers of limestone would react once the lower parts of the bordering mountains were covered with water.

Construction work began in 1956 and the dam was finally completed in 1960. Filling was initiated in February of that year, and within a month water had risen 400 feet above the floor of the valley. At that point some movement was observed on the neighboring slopes. By October, when the water level had reached 500 feet, it became clear that soil was moving downhill at more than an inch a day, and later that month a huge mass of rock and earth showed signs of movement. There were fears that a major landslide could occur. A month later a mass of surface material amounting to a million cubic yards crashed down into the lake in less than 10 minutes. The level of the reservoir was immediately reduced to 400 feet.

It was clear that the amount of water and rate of filling affected the stability of rocks and soils on both sides of the valley. The south side of the valley seemed to be less stable than the north, but nothing could be done to correct this weakness. There were no effective ways of stopping a slide if one occurred. It was decided to take preventive action in two ways: varying the level of water in the dam, and building drainage tunnels to divert water from the slide area. At the same time careful measurements were made to find out the relationship between the level of water and the incidence of slides so that controlled landslides could be initiated. As long as the mass and speed of a slide were within given limits, there was no risk of creating a wave to overspill the dam and cause destruction among the settlements below.

Over the following two years a carefully planned pattern of filling the dam while checking soil and rock creep on the mountains on both sides of the valley was followed. As long as this creep was within safe limits, the filling of the dam reservoir continued. By the beginning of October 1963 the engineering staff at the site felt confident they could raise the water to 700 feet without triggering a serious increase of soil creep on the valley slopes. The thousands of villagers who lived in the valley below were far from being comfortable with all of this experimentation. The 900-foot-high edifice looming above them gave rise to many sleepless nights. In the three years following completion in 1960 there was a pervasive element of fear in all the homes and villages beneath the dam.

Some of the technicians who were involved in maintenance work on the project expressed comparable concern over the danger from landslides. They pointed out that the mountainsides were dry and inclined to crumble because there was no vegetation to hold the soil in place. Their fear was that a heavy storm or some large rocks hitting the reservoir could cause water to cascade over the top of the dam into the valley. Even a small overflow of water from such a height could be disastrous. Concerns mounted as slippage along the face of the mountains increased in early October. They sent a statement of their concerns to the relevant government department in Rome. While waiting for a response there was a sudden change in the weather. After weeks of dry, hot conditions there came heavy rain and high wind. Meanwhile the report to the government had to pass through various levels of bureaucracy, so it did not reach Rome until early October 1962.

The environment changed dramatically in the first week of October as the ground became saturated with water. Groundwater rose, saturating the surface and decreasing its strength. This was especially relevant for sedimentary rock, in which everything is held together by the historical cohesion forged over time between the layers. By October 8, the day before the disaster, movements of surface materials reached the alarming rate of 16 inches a day. The last measurements on October 9 indicated double that rate in some areas. Villagers noted that animals must have sensed the danger because they began moving away. Engineers became alarmed and attempted to lower the water level, but even as they did so the reservoir level continued to rise as the slow creep of material displaced the water.

In investigations subsequent to the disaster, it was discovered that the layers of sedimentary rock on either side of the valley could easily be penetrated by water if the pressure were high enough. Water pressure did increase as the dam was being filled, and as water was pressed into the limestone it found a sliding surface in numerous thin clay layers. These clay layers lost their cohesion and thus their resistance to shearing. Added to these findings was the basic conclusion that the area was unsuitable for a dam. Had intensive geological studies been conducted at the beginning, particularly, testing the susceptibility of the surface materials to water, different decisions might have been made before construction began.

CONSEQUENCES

Late in the evening of October 9 a large block of rock, soil, and debris a mile wide, more than a mile long, and about a thousand feet thick roared down the mountainside at 70 miles per hour into the reservoir, displacing huge quantities of water. A gigantic, 300-foot-high wave was generated and this mass of water, like a tsunami, swept over the top of the dam, down into the valley, and into the Piave River northward and southward. The dam remained in place but destruction in the valley was catastrophic.

The water rushed down the valley, destroying everything in its path. Village after village and one home after another disappeared, leaving behind a mass of mud mixed with bodies and bits of building materials. Some people farther down the valley heard the sound of the approaching wave as if it were a tornado and managed to get out of its way in time. They knew at once what had happened. Longarone, the largest community, experienced the greatest amount of damage because it stood on the opposite side of the Piave River at right angles to the approaching wave. It was completely demolished and its population of 2,600 made up almost all the fatalities.

A day later, onlookers compared the scene with the ruins of Pompeii. The destruction of that city by volcanic eruption more than 2,000 years earlier was the only event they could recall that caused comparable devastation. In Longarone, as in all the other affected villages along the Piave River valley, population records were lost because the official buildings were destroyed, so it took some time to assess the loss of life. Gradually, as survivors met, the full toll became clear. The dam was never used again as a source of electricity. Even as the enormous scale of the tragedy was being grasped there was more terror. On October 15 there was another slide of rock into the reservoir. This time authorities were fully prepared. An evacuation plan was in place and buses quickly carried people to safety.

CLEANUP

The inhabitants of Longarone waited for almost three years before their homes were rebuilt. They had to fight local authorities in the process because of the loss of records. Throughout the valley all the good agricultural areas were either washed away or covered with debris. Some villages were abandoned, and their surviving residents went to live with relatives in Trieste, Milan, or Turin. In a few villages people returned within the following two years and rebuilt. A new community was created for some, a short distance away from the Vaiont River. It was named Vajont.

Questions were raised over the delay in the report's reaching Rome. Why had it not reached government inspectors sooner? Why were the warning signs that appeared early on in the process of construction not given more weight? In the end, nine men were accused of gross negligence and a trial date set. Five years later, on the night before their trial, the leader of the nine took his own life. The others received sentences ranging from fines to prison terms. One final outcome was that authorities launched a series of investigations into all dams in Italy's alpine region.

SELECTED READINGS

Chadwick, Wallace L. *Report to U.S. Department of Interior on Failure of Teton Dam*. Washington, D.C.: U.S. Government Printing Office, 1976.

Geiger, Charles W. "St. Francis Dam Collapsed." *Scientific American*, June 1928.

Kiersch, G. A. "Vaiont Reservoir Disaster." *Civil Engineering* 34 (March 1964): 32–39.

Leslie, Margaret. *Rivers in the Desert: William Mulholland and the Inventing of Los Angeles*. New York: HarperCollins, 1993.

3

Government Actions

Atomic Bombing
Hiroshima, Japan
August 6, 1945

I noted in the introduction to this book that it does not include case studies from war or health epidemics because events of these kinds are ongoing, not specific to a time and place. Hiroshima is an exceptional case. It was part of the warfare between the Allies and Japan during World War II, but it is included because its effects have influenced so many world events since 1945. The cold war between the Soviet Union and noncommunist countries was mainly focused on the dangers of nuclear weapons, and much of what we know today about dealing with nuclear power plant accidents came from experiences in Hiroshima.

Three weeks before the August 6, 1945, strike on Hiroshima, an atomic test bomb was set off in the United States to make sure that everything was working correctly. In spite of all the scientific planning that had gone into it, no one was quite sure how powerful the blast would be or how widespread the damage it would cause. Measurements taken within one second of the explosion told the story of the bomb's extraordinary destructive capacity. Within 6 milliseconds the blast had wiped out everything within 500 feet, and at the end of 50 milliseconds it had extended the devastated area fourfold. Hiroshima in August 1945 was a city of 245,000 people, about 100,000 fewer than its population at the beginning of the

A dense column of smoke rises more than 60,000 feet into the air over the Japanese port of Nagasaki, the result of an atomic bomb, August 9, 1945. (© Library of Congress)

war because many children and others had been evacuated to the countryside for safety. Almost all the dwellings were of wood construction, half of them single story and half one-and-a-half stories. Fire-fighting equipment was antiquated.

CAUSES

Why was the atomic bomb dropped on Hiroshima? There are many answers to this question because there were many people involved in the decision. The best estimate is related to strategic plans. The United States wanted to force Japan's surrender as quickly as possible in order to reduce American casualties. Alongside that concern was a desire to prevent the Soviet Union from becoming involved in the conquest of Japan. By July 1945, the Soviet Union had attacked and occupied some islands north of Japan. Although allied with the United States at the time, the dictatorship of the Soviet Union was not trusted by the U.S. Many scientists were strongly opposed to using the bomb on civilian targets, and President Truman also expressed deep concern about the same thing. The overarching consideration that settled the question was the saving of many thousands of American lives that would certainly have been lost without the bomb.

Hiroshima was chosen as the target because its size and the nature of the surrounding terrain made it suitable for investigating the destructive capabilities of the bomb. Other reasons were the concentration of military installations, munitions factories, and troops. Because no bombing had previously been carried out there, the advantage of surprise existed, an important consideration in a critical mission. *Enola Gay*, the B-29 plane that carried the atomic bomb, left the island of Tinian shortly before 3:00 A.M. on August 6 and arrived over Hiroshima at 9:00 A.M. The bomb was released and detonated in the air above the center of the city. The explosion was so powerful that the B-29, 12 miles away when the bomb went off, shook violently for several minutes.

CONSEQUENCES

Local resources and services in Hiroshima were devastated by the bomb and the city had to wait for help from elsewhere. For an area of 4.5 square miles around ground zero, the point on the ground immediately beneath the bomb explosion, every living thing was destroyed. The destruction unleashed on the city was total. Seventy percent of all buildings and 80,000 people were obliterated in an instant. Japanese authorities learned for the first time of the awful power of this new bomb. Emperor Hirohito decided to end the war as soon as possible, but in Japan's authoritarian society, with so much military control, that decision could not be implemented quickly.

In an atom bomb explosion extremely high pressures and equally high temperatures are present, far greater than are ever experienced in industrial enterprises. The metal framework of one building—the city hall—was all that was left standing. Beyond the vaporized zone the supersonic blast of air and heat, releasing millions of degrees of heat, destroyed everything. People standing 10 or more miles away were burned right through their skins. They died either immediately or soon afterward. Iron, stone, and roof tiles were twisted out of shape. Clothing, railway ties, and trees instantly ignited. At 800 miles an hour the hot air created huge swirls of wind that circled back into the city to fill the vacuum initially created. Whatever remained from the initial blow was destroyed then. Black, radioactive rain followed soon afterward as dirt, water, and debris that had been sucked up by the mushroom cloud came down.

CLEANUP

The harm done to humans was quite a different story. In 1945 no one knew much about nuclear radiation or the diseases it would generate. Many of the people who survived the bombing received heavy doses of radiation. In the months that followed, large numbers of these people died from various illnesses, and many more succumbed in the years that followed. By December 1945 the death toll had risen to 140,000. Leukemia, cataracts, thyroid cancer, destruction of reproductive organs in both women and men, and lung cancer were among the diseases experienced. Over the years since then casualty numbers have continued to mount, but statistics are difficult to collect because of other health complications independent of those caused by radiation.

Today there is little to see in Hiroshima of the destruction that destroyed the city in 1945. The city was rebuilt and is now a thriving business and manufacturing center. A centrally located museum has full documentation of all that happened, including photographs of the event and of people's injuries at the time of the bomb as well as at different stages over the succeeding years. The only part of the original city still remaining—and forming part of the museum—are the city hall's foundation and the bands of steel that once framed its roof. Every year, from all over the world, visitors in large numbers come to Hiroshima. It is a major tourist attraction, and tourism is a significant source of income for the city. The visitors want to know how a large urban area survives an atomic bomb attack and what people today think about what happened in August 1945.

Atomic Bomb Testing
Bikini Atoll, Marshall Islands
1946

Bikini Atoll is one of a group of atolls that compose the Marshall Islands, which was made a U.S. protectorate under the United Nations in the years following World War II. These islands are located about halfway between Hawaii and Australia and stretch over the Pacific Ocean for hundreds of thousands of square miles. The first Europeans to reach the area, more than 300 years ago, were the Spanish. For most of the succeeding 250 years there was little continuing contact with Western nations. Trade in copra oil from coconuts was carried on, but the islanders remained isolated, free to live in their closely knit society strengthened by extended family ties and local traditions.

All this changed dramatically after World War I when Japan was mandated by the victorious nations to govern the islands. A military buildup began on Kwajalein Atoll, 200 miles to the south of Bikini, and this island became headquarters for Japan's armed forces in the Marshall Islands. A watchtower was built to guard against possible invasions. This atoll remained a fortified military position throughout World War II and was finally captured by U.S. forces in 1944 after a terrifying and costly conflict.

After World War II President Truman directed the U.S. Navy to investigate the effects of atomic bombs on U.S. warships. Bikini Atoll was chosen for this project for several reasons. It was located away from regular air and sea routes, had a good-sized lagoon, and nearby were a few large islands that could serve as observing stations. There was good access to the lagoon through wide channels, and a shallow area a few miles offshore was a suitable site for anchoring the target ships. The islands of Kwajalein and Enewetak were close enough to serve as bases for aircraft.

In February 1946 the Bikinians—all 167 of them—were asked if they would be willing to leave temporarily so that the United States could begin testing atomic bombs. The U.S. governor assured them that this move would be for the good of mankind and would help to end all world wars. After much sorrowful deliberation, King Juda, the Bikinian leader, announced, "We will go believing that everything is in the hands of God." The Bikinians were sent to Rongerik Atoll, 125 miles to the east, a place about one-sixth the size of Bikini that was traditionally regarded by the Bikinians as unlivable because of inadequate resources of water and food. There was also a deeply felt conviction that Rongerik was inhabited by evil spirits.

The Bikinians were given food supplies for several weeks and then left to fend for themselves. Soon after these provisions ran out the islanders' worst fears began to surface as they were unable to find adequate local food supplies of the kind they were accustomed to— coconuts and fish. A rash of serious illnesses broke out late in 1946, perhaps due to lack of food, and a fire that damaged a large number of coconut trees reduced them to near starvation. They begged the navy to let them return to Bikini. Instead they were moved to Ujelang Atoll early in 1947, but something went wrong with that plan and their odyssey took them to yet another group of islands, Kwajalein, where they were housed in tents on a strip of grass beside the airport.

CAUSES

The reason for these migrations was always the same: the operational needs of the atomic bomb tests. In 1948 there was another move and the Bikinians left Kwajalein for the island of Kili, far to the south of the Marshall Islands. While the islanders struggled to set up their new community on Kili, Bikini continued to be irradiated with a steady succession of bombs. The first two, each of the same power that had devastated Hiroshima in 1945, were detonated in the air on July 2, 1946. They were followed by more than 20 additional blasts between the years 1946 and 1958, some on the ground, some above it. A fleet of 100 warships along with 40,000 men and women, hundreds of animals, and 20 tons of photographic equipment were involved in these activities.

In 1954 the new and vastly more dangerous hydrogen bomb, a thousand times more powerful than the atomic bomb that destroyed Hiroshima, was detonated on the ground. Weather forecasting was very different in 1954 from what it is today. There were no satellites and no computers. Furthermore, little historical data was available for Bikini so it was very difficult to predict how winds might change in the short term. The enormous amount of planning and arranging that went into a bomb test, especially with this new type of bomb, necessitated a final decision days ahead of the blast. Winds were favorable all day right up to eight hours before blast time, but then they changed. The test went forward in spite of this outcome, and everyone knew that radiation would be blown onto some inhabited islands.

The site was the surface of the reef in the northwestern corner of Bikini Island. The area was illuminated by a huge and expanding flash of blinding light. A raging fireball of intense heat measuring millions of degrees shot skyward at a rate of 300 miles per hour. Within minutes the monstrous cloud, filled with nuclear debris, shot up more than 20 miles and generated winds of hundreds of miles per hour. These fiery gusts blasted the surrounding islands and stripped the branches and coconuts from the trees. Millions of tons of sand, coral, and plant and sea life from Bikini's reef and the surrounding lagoon waters were sent high into the air. The force of the explosion far exceeded the expectations of observing scientists. They expected a force of 5 megatons and they experienced 15. It was the most powerful bomb ever exploded by the United States. Fifty thousand square miles of land and sea were contaminated.

Ships stationed about 40 miles east and south of Bikini in positions enabling them to monitor the test detected the eastward movement of the radioactive cloud from the huge blast. They recorded a steady increase in radiation levels so high that all men were sent below decks and all hatches and watertight doors were sealed. One-and-a-half hours after the explosion, twenty-three fishermen aboard a Japanese fishing vessel, the *Lucky Dragon*, watched in awe as a "gritty white ash" began to fall on them. The men aboard the ship were oblivious to the fact that the ash was the fallout from a hydrogen bomb test. Shortly after being exposed to this fallout their skin began to itch and they experienced nausea and vomiting. One man died.

CONSEQUENCES

Three to four hours after the blast, the same white, snowlike ash began to fall from the sky onto the 64 people living on Rongelap Atoll, about 125 miles east of Bikini. Not understanding what had happened, they watched as two suns rose that morning and observed with amazement as the radioactive dust soon formed a two-inch-deep layer on their island, turning their drinking water into a brackish yellow liquid. Children played in the fallout. Their mothers watched in horror as night came and they began to show the physical signs of exposure. There was severe vomiting along with diarrhea and hair loss. The islanders fell into a state of terrified panic. Two days later they were finally taken to Kwajalein for medical treatment.

Throughout all of the tests, eagerness to examine the experimental animals and the rocks and water areas around the explosion sites, coupled with general ignorance about radiation dangers, led to huge risks being taken in the early days of the tests. It had been less than a year since the bomb was dropped on Hiroshima and as yet there was little evidence of continuing radiation effects. Scientists, political leaders, and military men boarded ships within hours of an explosion to visit the sites of the bomb tests. They examined dead fish and animals and carried them back to their observation stations for examination. During their moments of free time they swam in water less than a hundred yards from the site of the explosions.

By the end of 1946 several men became ill and had to be taken to the naval hospital in Hawaii. For most of them, medical discharges were given in due course and they went home. In the late 1950s the tests ended, and everyone returned home or to their military bases. It was not long before the first of numerous cancers appeared and the reality of long-term radiation damage was known. Evidence from Japan of similar outcomes confirmed the tragic reality. While the exact figures were never known because of other health factors, it was concluded that thousands of people who were involved in the Bikini tests died prematurely from various radiation-induced diseases.

About 12 years after the series of bomb tests was completed, U.S. government agencies began to consider returning the Bikinians to their homelands in compliance with the original promise in 1946. Specialists measured radiation levels on Bikini Atoll, and it was considered safe. One report from the Atomic Energy Commission (AEC) went so far as to say, "Well water could be used safely by the natives upon their return to Bikini. It appears that radioactivity in the drinking water may be ignored from a radiological safety standpoint. The exposures of radiation that would result from the repatriation of the Bikini people do not offer a significant threat to their health and safety." Accordingly, in June 1968, the 540 Bikinians living on Kili and other islands at that time returned to their homeland.

For seven years there was little indication of any problem. The population of Bikini slowly increased. Then in 1975, during regular monitoring, radiological tests discovered higher levels of radioactivity than was originally thought. U.S. Department of Interior officials stated that "Bikini appears to be questionable as to safety," and an additional report pointed out that some water wells on Bikini Island were also too contaminated with radioactivity to be used as sources for drinking water. A couple of months later the AEC, on review of the scientists' data, decided that the local foods grown on Bikini Island—pandanus, breadfruit, and coconut crabs—were also too radioactive for human consumption.

CLEANUP

After the Bikinians decided to take action, they filed a lawsuit in U.S. federal court demanding that a complete scientific survey of Bikini and the northern Marshall Islands be conducted. The lawsuit stated that the United States had used highly sophisticated and technical radiation detection equipment at Enewetak Atoll but had refused to employ it at Bikini. More than three years of bureaucratic squabbles among the U.S. Departments of State, Interior, and Energy over costs and responsibility for the survey delayed any action on its implementation. The Bikinians, unaware of the severity of the radiological danger, remained on their contaminated island.

In April 1978 medical examinations performed by U.S. physicians revealed radiation levels in many of the 139 people on Bikini to be well above maximum permissible level. The next month the U.S. Department of Interior announced plans to move the people from Bikini "within 75 to

90 days." Thus, in September 1978 people were once again evacuated, and this time it was a final departure. It would be impossible for them to live again in their homeland as they had before the tests. They saw Bikini again only once, in 1988, when they were brought back to witness the beginnings of a long-term project, the decontamination of the soils surrounding the lagoon. Some day, when as yet unknown because of the enormity of the damage, that project may be completed so that a future generation of Bikinians can live there.

After the people of Bikini were removed from their atoll for a second time, the U.S. government established a $6 million trust, the Hawaiian Trust Fund for the People of Bikini. A second grant of money, $20 million, was given to the Bikinians in 1982 to help them whenever they could return to their homeland. In 1997 there was a third grant, $90 million, this time for the cleanup of Bikini and Eneu, two of the islands of Bikini Atoll. The total value of the fund by the year 2000 was $130 million, and 90 of the original 167 Bikinians who left in 1946 were still alive at that time.

In March 1998 the International Atomic Energy Agency (IAEA) presented its final report on radiological conditions at Bikini Atoll. It concluded that, on the basis of the amount and quality of information now available, no further testing would be necessary. The Bikinians should not be allowed to return to their homeland permanently and eat locally grown food until remedial measures are carried out. However if the food they consume is imported, as is presently the case in a number of sites on the atoll where fishing in the lagoon and diving-tourist enterprises on the big sunken ships are in operation, there are no dangers associated with temporary occupancy.

The IAEA is sure that it is safe to walk on all of the Bikini Atoll islands. While the residual radioactivity is still too high for growing crops, it is not hazardous to human health. The air, land surface, lagoon water, and drinking water are all safe. There is no radiological risk in visiting the lagoon or the islands. The nuclear weapon tests have left practically no cesium in marine life. The cesium deposited in the lagoon was dispersed by the ocean.

One measure under consideration by the IAEA for ensuring the return of the islanders with freedom for them to eat locally grown food was based on using a potassium-based fertilizer. This would be spread on all areas of Bikini Island. Soil beneath and around homes would be removed and replaced with crushed coral. Because Bikini Island soil is extremely deficient in potassium, it has been found that plants will choose the coral when it is available rather than radiated minerals. The problem with this approach is its short life. It will work for four or five years only before repeated applications of the fertilizer are needed.

The Bikinians are wary of any short-term measures. They have been badly hurt already by the temporary arrangements that brought them close to starvation. They favor the IAEA's soil-removal approach whereby the top 15 inches of soil are removed. This would eliminate the danger of radiation but would be environmentally costly. The fertile topsoil supports the tree crops, which are a major food resource. Nevertheless the Bikinians continue to campaign for soil removal for all 23 islands of the Bikini Atoll. They would like to see it happen first on Bikini Island, with the excavated soil being used to build a causeway to a neighboring island that is presently accessible only at low tide.

The enormous ignorance that prevailed in 1946 about the dangers of radiation from atomic bombs created a lethal atmosphere around the Bikini test area. Some of the tragic effects of this were immediately evident, but others came to light gradually over the years. Hundreds of military personnel who were involved in the tests suffered from radiation diseases of various kinds. Today the dangers from radiation are well known and protective measures are in place. Of the more than 200 ships that were brought to Bikini for the tests, 10 of the largest, including the former aircraft carrier USS *Saratoga*, still lie in the Bikini lagoon. They have been decontaminated and serve as popular sites for diving tourism.

Landscape Devastation
Ukraine, Soviet Union
1932

In the aftermath of the Bolshevik Revolution of 1917, Russian peasants obtained additional land from the owners of the large farms and Vladimir Ilyich Lenin, the Communist leader, encouraged them to do this. He saw a period of small-scale free enterprise as a useful intermediate stage on the path to dictatorship. This arrangement went on for some time, and farmers continued to work their land for profit. After Lenin died, Joseph Stalin came to power and within a few years decided it was time to abolish all private ownership of land and establish collective farms. This decision was part of a much bigger plan to double the nation's industrial output. To accomplish it, Stalin felt he had to secure complete control of agricultural production and thus guarantee adequate food supplies for the factory workers.

The focus of his plan was the Ukraine, where the best agricultural land of the nation was found and where he soon encountered the strongest opposition. The small farmers were determined to retain possession of their farms, and when they saw that Stalin was planning public ownership of all farms they resisted. Their first move was to kill off all their livestock for food and hold back as much of their grain crops as they could. In less than a year these moves began to starve the cities of their food supplies, and Stalin's drive for industrialization was threatened. Twelve million new workers had already been added to the industrial enterprises around Moscow over the course of two years. Consequently, more food was needed just when existing supplies were reduced. Stalin became desperate and decided to act decisively against the Ukrainian farmers.

CAUSES

Certain philosophical approaches to national development were widely held by the Communists and it is helpful to note these in order to understand the reasons for the total landscape devastation that followed Stalin's actions. For one thing, they were convinced that government control of all agricultural and industrial production was more efficient than private control. They also believed that they could modify natural environments, such as land and soils, to make them produce what the government needed. Trofim Lysenko, a biologist and geneticist, gave support to this kind of thinking by claiming that he could create better crops and animals by cross-fertilization. He even suggested it would be all right for the state to decide where dairy farming should be conducted and where wheat ought to be grown without regard to the nature of the environments in the places selected. His theories were unscientific but they were very popular with Stalin.

The total collectivization of the small Ukrainian farms looked like a military campaign as thousands of soldiers entered the region. The slightest opposition on the part of the farmers was met by death or banishment to Siberia. The existing allocation of grain for the government, already so large that what was left barely prevented starvation, was suddenly doubled. This caused widespread famine. Because all grain harvested was closely supervised by soldiers and the government quota strictly guarded, the only way to avoid starvation was to hide grain secretly wherever possible, under the floorboards of homes or outside in underground caches. These moves led to the first of the destructive actions that would become commonplace. Soldiers raided homes in search of grain, tore up floors, broke down walls, and confiscated what had been hidden outdoors. The demolition of the farmers' homes did not matter once grain was discovered. The entire family would soon be dead or gone. Thousands of homes disappeared in this way.

Old rivalries between Russia and the Ukraine came to the surface as opposition to collectivization proceeded. The Ukraine had cultural symbols, mainly in churches, that were different from those in Russia, and Stalin felt that these created a sense of independence and opposition to Russia. A deliberate scorched-earth policy was adopted toward them, and in the course of one year 250 churches and numerous monuments were demolished. Sometimes, as happened in Moscow during the November 1932 anniversary celebrations of the 1917 revolution, a single incident was enough to intensify the terror campaign against the Ukrainians.

In the course of the celebrations Stalin's young wife, Nadezhda Alliluyeva, criticized her husband publicly because of the widespread starvation in the Ukraine, especially the occurrence of cannibalism. Alliluyeva knew she had violated the code of secrecy about government business. Knowing the consequences, she committed suicide. At the same time Stalin arranged for the execution of every guest who had attended the celebrations. Immediately afterward he instituted a new passport system for the Ukraine to ensure that no one would be able to leave and report what was happening. Starving peasants who tried to leave in search of food were shot. Meanwhile, huge collective farms appeared, some as big as a quarter-million acres. Bureaucrats from the Communist Party's head office in Moscow were put in charge of them, deciding what to plant and where, what machinery to buy, and how to use it, all without any expertise. The only sources of knowledge for this work were either dead or in Siberia.

CONSEQUENCES

The intensive use of the rich black soils of the Ukraine under years of pressure from Moscow to maximize wheat production had already taken its toll, reducing soil nutrients and compacting the ground. There was now the added loss of organic fertilizer because almost all the farm animals were gone, further deteriorating the soil and making it vulnerable to wind erosion. In other wheat-growing countries fallow years and strip farming compensate for these limitations. In spite of all his diabolical approaches to maximizing grain production, Stalin soon found that the whole system was unworkable. The tractors he brought to speed up farm work broke down constantly, and soldiers could not fix them. He was unable to revert to the system formerly in use by the small farmers because very few draft horses were still alive.

There was no one around who had experience when weather conditions required changes in practice. The end result was that grain supplies for the industrial centers remained inadequate. This time Stalin had only his own soldiers to blame. In the course of 12 violent months seven million people had died through either military executions or starvation. Students from the Soviet School of Mines in Moscow and other colleges were sent to provide emergency help to the soldiers in the collectives because they had some knowledge of the region. Each group of students was sent to particular villages, but as they traveled through the Ukraine they noticed that there were few people anywhere. Sometimes on arriving at a place recorded on their maps as a village, expecting to be housed and fed, they found nothing but a few bricks and weeds.

CLEANUP

The 1933 grain harvest was worse than the 1932 one. Reluctantly, Stalin had to admit defeat. He introduced an incentive program for those Ukrainians who were still alive and healthy enough to take advantage of it. Anyone who worked on a collective farm was permitted to have a plot of land on which crops for private use could be grown. Over the year that followed this concession, productivity of the private plots was many times greater than that of the large collectives. The outcome of the incentive program as far as government needs were concerned, however, was never known. Within a very few years the country was embroiled in World War II and other things took priority.

The Western world was largely unaware of the Ukrainian horrors at the time, and Stalin made sure they remained unaware. Correspondents from major U.S. newspapers were given privileges if they reported favorably on what was going on, and several did. Because every report was censored they felt that this was the only way they could do their work. Stalin went further than influencing newspaper readers. He designed model collectives where everyone seemed well fed and happy in their work, then invited influential leaders from the West to visit them. Edouard Herriot, twice premier of France and a socialist, spent five days in the Ukraine and stated that there was no famine there. John Maynard Keynes, one of the world's greatest economists of his time and an expert on Russian agriculture, visited the Ukraine and told everyone in Great Britain when he returned that reports of famines were totally unfounded. Stalin was helped in his efforts to deceive by the West's preoccupation with the Holocaust of Nazi Germany. A belated recognition of Stalin's landscape holocaust came in 1984 when President Ronald Reagan declared November 4 of that year "A day of commemoration for the Great Ukrainian Famine of 1932 and 1933."

Arsenal Explosion
Military Arsenal, Morris County, New Jersey
July 10, 1926

The U.S. Navy's ammunition depot was located about 30 miles west of the Hudson River in New Jersey, near the village of Dover and south of Lake Denmark. The entire supply of munitions was stored there, and ships came up the Hudson River for supplies that were then transferred to them by rail from the depot. In 1926 the fleet was concentrated on the Atlantic Coast, so this location was accessible at all times. Thirty buildings in the depot complex were devoted to the manufacture of the army's smokeless powder. In all there were 180 buildings in the complex. Because of the huge quantities of explosives on hand at any time and the frequency of summer thunderstorms in this part of New Jersey, a lightning-arresting system had recently been installed.

CAUSES

On July 10, at about five o'clock in the afternoon, a storage unit that housed a million pounds of high-intensity TNT was hit by lightning. That should not have been a problem as the lightning rod was within a few feet of that location. For some unknown reason it failed to work. Within minutes a column of black smoke appeared, followed by a small fire. Immediately an alarm rang out across the entire complex. The commander of the depot, who was half a mile away in his home, received a call from the officer on duty at the same moment. He knew what was stored next to the small fire, and in less than a minute he was running to his car, hoping to reach it before the inevitable explosion took place.

His wife was delayed for a few minutes. As she ran out of the house seeking refuge in a nearby forested area, the first blast arrived and the house crumbled like a pack of cards. She was showered with shards of glass and bits of plaster. A huge white flash was followed by a series of deafening roars.

CONSEQUENCES

A wall of flame swept over the depot as case after case of ammunition exploded. The pressure wave was sufficient in most cases to demolish building after building, and whatever was left was devoured by the flames. From the moment of the first explosion, flames continued to burn for 10 more hours, destroying both the buildings in the depot and the surrounding countryside. The entire area within half a mile of the depot was flattened. Of the

180 buildings that composed the arsenal, only 16 survived. Residents as far away as 50 miles heard and to some extent felt the blasts. When it was all over, the damage added up to more than $150 million. Thirty lives had been lost and 200 others injured. Visitors to the area who happened to be driving close to the navy's arsenal as explosions were occurring encountered all kinds of debris on the roads, including live rounds of six-inch shells.

The countryside was torn up around the arsenal, leaving it like an old-style battlefield. Deep craters were torn out of the sandy soil. Steel girders were thrown a mile from the blast site. A large piece of a charred wooden beam was found on a farm three-and-a-half miles from the blast site. One of the buildings inside the complex was a medical station fully equipped with medicines and communications equipment. It was destroyed before it was ever used.

CLEANUP

New Jersey's governor together with the state's two senators visited the site on July 11 and immediately raised the question of the arsenal's location. Was it right to have it so close to populated areas? Was it wise to store the entire stock of navy munitions in one place? What would have happened if the country were at war and an accident of this kind happened? The general public was very supportive of the need for a better site as well as the desirability of having several sites.

The navy was blamed for not taking adequate safety precautions. The lightning-arresting system was inadequate, and the high concentration of explosive material in so small a space was unwise. Reaction was strong from the U.S. government, both for the cost of the tragedy and for the dangerous position in which it placed the navy. Two years after the explosion the Department of Defense's Explosives Safety Board was formed. It was instructed to provide oversight on all aspects of explosives, including maintenance, transportation, and storage. Never again would so high a concentration of munitions be found in one place. Arsenals were subsequently placed in low population density areas across the country.

Shipping Explosion
Halifax, Nova Scotia, Canada
December 6, 1917

At the peak of World War I, when the United States was fully engaged in the European conflict, the port of Halifax, in Nova Scotia, Canada, was a key link in the supply of munitions to the armies in France. It was on the shortest path from New York to Europe and was the last port of call for ships on the North American mainland. The docks at Halifax became a transshipment center for supplies and personnel about to be sent to Europe. Huge storage bins and temporary accommodations had been built beside the docks to provide for the materials and people waiting there. The whole port looked like an industrial town.

CAUSES

The *Mont Blanc*, a Belgian ship that just arrived from New York, was about to enter Halifax harbor. It was fully loaded with munitions for the armies in Europe and was stopping in Halifax to load coal. In 1917 coal was the engine fuel used in ships, and Halifax was the last place to stock up on this fuel in case of delays on the voyage. Any ship carrying munitions normally flew a red warning flag to alert other ships to the danger. On this occasion, because it was wartime the ship's captain chose not to fly it. Everyone knew that German submarines prowled the western Atlantic, and a munitions ship would be an ideal target for them.

The entrance to Halifax harbor was very narrow, about half a mile wide, and all ships were urged to take extra care when entering or leaving. Traffic was heavy in the narrow channel on December 6 because in winter the number of hours of daylight is at a minimum, and in a northern country such as Canada that minimum is very small in December. Given the type of navigational equipment available in 1917, every ship wanted to get out of the harbor before dark. One Norwegian ship, the *Imo*, was about to leave port as the *Mont Blanc* was arriving. The *Imo* had been delayed and its captain was anxious to get away. His ship entered the narrow channel from the west as the *Mont Blanc* entered it from the east (see figure 3.1).

As the two ships approached each other, Captain Le Medec of the *Mont Blanc* signaled to the *Imo* that he wanted the ships to pass on their port sides (the left sides of the ships when looking in the direction of travel). According to standard rules of navigation that is the proper mode of passing when ships meet, and any other arrangement is permissible only with mutual agreement. The *Imo*'s captain replied that he would prefer a starboard passing (that is, on the ships' right sides), presumably because he thought it was a quicker route. Unfortunately the *Imo*'s captain did not wait for a reply before taking action. Le Medec suddenly became aware that the rules of the road were not being followed. He tried to swing his ship around in a desperate effort to avoid a collision, but in the narrow channel and with so little time available it was not possible.

Figure 3.1
Converging Paths toward the Halifax Shipping Explosion (Paul Giesbrecht)

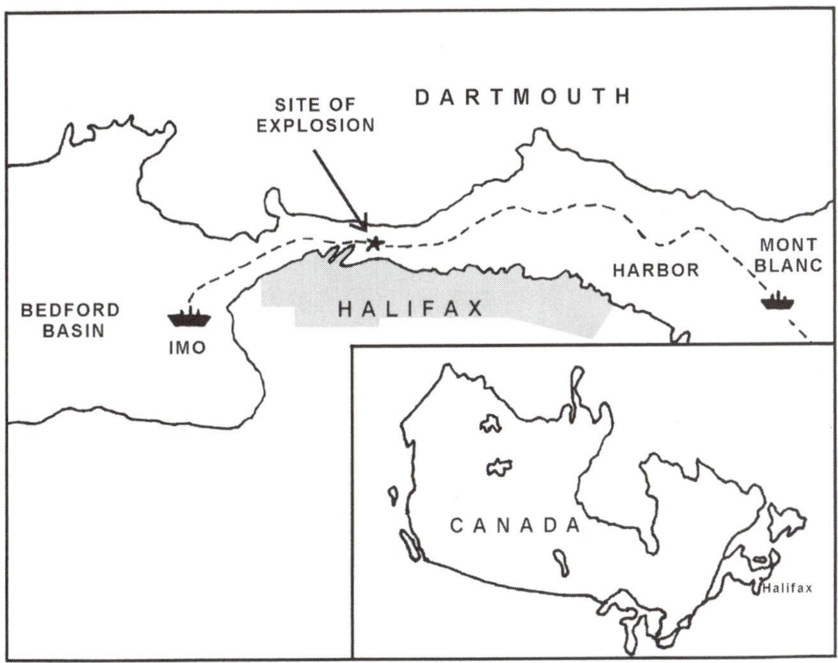

The *Imo* hit the *Mont Blanc* near its front end, very close to a hold storing huge quantities of TNT. A drum of benzol was broken open in the collision and a spark ignited it. As the liquid ran down into the hold the fire spread to the TNT. The *Imo*'s captain, unaware of the cargo on board the *Mont Blanc*, backed his ship away and beached it. Captain Le Medec knew what was about to happen. TNT needs either heat or extreme pressure to explode. In the moments before things reached the critical level of heat the ship drifted toward a pier on the south side. A telegraphist was on duty there. He watched the ship come closer and closer and he knew about its lethal cargo. He stayed at his post, sending out warning signals, and just before the explosion he shouted a good-bye. His body was found later that evening.

CONSEQUENCES

The power of the explosion swept away churches, factories, and every other kind of building. The *Mont Blanc* disappeared in a cloud of smoke, and a blast of air and debris rose one mile high. Nothing was left standing.

Fires broke out all over Halifax and shells and other ammunition rained down on people, some of it exploding in the process. Terror-stricken men, women, and children, all of them covered with black soot and bleeding from flying pieces of glass, were seen everywhere. Everything within a one-mile radius was devastated. Some bodies were thrown half a mile onto the shore. A 30 foot tsunami erupted and, carrying rocks it scooped up from the seabed, destroyed all the piers together with their contents.

Large numbers of military personnel and supplies were assembled at the harbor, awaiting transportation to Europe. Two further complications made circumstances specially difficult for victims. The worst blizzard in living memory had just hit Halifax and the air temperature was well below freezing. Many victims who were trapped in buildings froze to death. The second complication was a forced evacuation because of the danger of fire at the munition stores close to the piers.

Out of a total population of 50,000, 1,600 people lost their lives and a further 9,000 were injured. Six thousand people lost their homes. The violence of the explosion, from today's vantage point, can be compared with that of an atomic bomb, but without the deadly radiation associated with the latter. Windows were shattered a hundred miles away. Damage amounted to $35 million, an enormous sum at that time. Subsequent investigations concluded that the captains of both ships were to blame for what had happened, but only minor penalties were given.

CLEANUP

Halifax was completely rebuilt, and today a museum displays the details and artifacts from the explosion.

SELECTED READINGS

Christman, Albert B. *Target Hiroshima*. Annapolis, Md.: Naval Institute Press, 1998.

Conquest, Robert. *The Harvest of Sorrow: Soviet Collectivization and the Terror Famine*. Edmonton, Canada: University of Alberta Press, 1986.

Kitz, Janet F. *Survivors: The Children of the Halifax Explosion*. Halifax, Canada: Nimbus Publishing, 1992.

Weisgall, Jonathan, M. *Operation Crossroads: The Atomic Tests at Bikini Atoll*. Annapolis, Md.: Naval Institute Press, 1994.

4

Industrial Explosions

Ammonium Nitrate Explosion
Oppau, Germany
September 21, 1921

The Badische Anilin and Soda Fabrik Company opened a factory in Oppau, Germany, in 1911 to produce chemical products, including agricultural fertilizers. Oppau was a small community of about 6,500 people not far from Mannheim in western Germany. A new process had just been developed by German chemical engineers to produce these fertilizers from ammonium sulfate. The traditional natural supply from organic waste was in short supply, so this was a timely substitute for Germany's agricultural industry. At the time the country was prosperous, with an expanding economy, and it needed a stronger agricultural base. Because farmers use most fertilizers in springtime, the factory stockpiled large quantities throughout the rest of the year in preparation for their spring sales. A large number of 60-foot-high silos were built to handle the storage.

Other products, such as explosives, can also be produced from ammonium, and the factory in Oppau served the nation well in this respect during World War I. It manufactured explosives, but it was also responsible for making the poison gas that German soldiers used in the course of the war. As a result, a great deal of secrecy surrounded Oppau during the war years, from 1914 to 1918. When Germany tried to cope with the devastation of the war in the wake of its defeat, it turned once again to the Badische Com-

French field kitchens feed children in Oppau, Germany, after the dye-works blast wrecked town, October 11, 1921. (© Library of Congress)

pany for fertilizers. The country's currency was so low in value that it was impossible to purchase the sulfur needed for the production of ammonium sulfate from other countries. The factory therefore switched to ammonium nitrate, an acceptable substitute, for its production of fertilizers. The new product was stored in the same 60-foot-high silos as before.

CAUSES

Ammonium sulfate, the product selected by the designers when the factory was opened in 1911, is not explosive, but ammonium nitrate is. This fact either was not known at the time, or no one told the workers in Oppau of it. Thus the workers carried on the same procedures as before in working with the new chemical product. To some extent the new fertilizer material was not explosive. It could only be ignited at very high temperatures or when subjected to great pressure. Workers soon discovered that it evinced a feature they had not encountered before: a tendency to attract moisture from the atmosphere and become sticky. Over time this characteristic caused the fertilizer at the bottom of the silos to bond into a solid mass, almost like concrete, under the pressure of the weight from above.

Loosening the ammonium nitrate in the silos for market became a challenging task. At first picks and shovels were used, but before long it was evident that much more powerful tools were needed. The Oppau workers decided to use low-grade dynamite. They bored holes in the nitrate, set the dynamite, and broke up the hard material into smaller pieces. Low-grade dynamite can be used to break up solid substances and was frequently employed on road and building construction sites to clear away large rock formations. It was a much more risky business to use it with ammonium nitrate, and it would never have been used at Oppau if workers had been familiar with the properties of the new substance. For a short time the method worked well and fertilizer material was being shipped out from the plant on time. Then, on the morning of September 21 the method failed. Too much pressure must have been placed on the ammonium nitrate. There was large blast and a huge fire at one silo, followed by a much bigger explosion as all the other silos erupted in flames.

CONSEQUENCES

A column of fire shot upward into the air a thousand feet, and everywhere was total destruction. A 50-foot-deep, 250-foot-wide crater was carved out beneath the factory. Much of the depression formed at that time can still be seen today. The entire town of Oppau disappeared. The blast was so strong that it shattered windows 40 miles away. The workers in the factory lived in three- and four-story apartment buildings within walking distance of the site, so they and their families had no chance to escape. Furthermore, the factory was isolated from other communities. There were no major medical facilities such as a hospital nearby—not even a doctor—so all the injured had to be taken to Mannheim, 40 miles away. Because of the widespread damage and large numbers of casualties there were numerous delays, and many people died before reaching a hospital. One unusual factor added to the confusion and delay. The roofs of the homes in Oppau were made of heavy clay tiles that lay on support beams but were not fixed to the roofs. As the explosive blast blew away the buildings, the tiles became dangerous missiles.

A great deal of suspicion accompanied the explosion, and it took a long time for newspapers to get accurate information. The Badische Company's earlier history as a manufacturer of poisonous gas gave rise to numerous exaggerated stories. When reporters and government officials were finally able to assess the damage, the death toll was found to be in excess of 500. The exact number was never known because modern investigative tools such as DNA analysis were unavailable at that time. More than 2,000 people were injured and many thousands left homeless.

CLEANUP

As in many industrial communities of the 1920s, one hard lesson was learned: that residential buildings must not be built close to factories. Even today, this rule is ignored in many developing countries because it is cheaper for people to live near their work. In developed societies such as Germany the lesson has been taken to heart. Oppau is now a residential area a considerable distance from industrial plants and known as Ludwigshafen-Oppau.

Ammonium Nitrate Explosion
Texas City, Texas
April 16, 1947

Texas City, about 30 miles southeast of Houston, Texas, on Galveston Bay, was a thriving oil port in the 1930s with a population of 6,000. Several oil refineries stood near the docks, and the shipping traffic was mainly occupied with crude oil and petrochemical products. With the onset of World War II in the first half of the 1940s and a rising demand for both aviation fuel and a range of synthetic chemicals, both the population of the city and its industrial capacity expanded dramatically. Production of oil-based products jumped fivefold in five years and the population more than doubled in the same time. By 1947 there was an aura of success and confidence as business continued to grow. In the aftermath of the war the dock capacity at Texas City was expanded in response to large new shipping demands.

A zoning law was passed in 1946 to establish which areas were to be devoted to industrial, residential, and institutional activities, and in each of these places safety precautions were given high priority. Gas and oil storage tanks were equipped with fire control systems and surrounded by dikes that would prevent spills reaching other buildings. There were good reasons for these precautions. Within a one-square-mile area beside the docks were six oil-company complexes, eleven warehouses, and several other installations and residential blocks. Fire-prevention experts assured the port authorities that only one-fifth of this area was in danger of a serious fire and that existing precautions would be adequate to cope with such an eventuality.

Firefighters and dock workers pulling fire hoses to the burning freighter *Grandcamp* in Texas City shortly before ammonium nitrate fertilizer on board exploded. (AP/Wide World Photos)

All of this thinking should have changed when large quantities of ammonium nitrate fertilizer began to be shipped from Texas City in 1946. Now, in the event of an accident, there was the possibility that a conflagration would affect the whole dock area, not just a fifth of it. Two conditions prevented the kind of new thinking that was required. The railway managers and ship's masters who were responsible for handling the nitrate operated independently of the other agencies, and they saw no reason for coordinating their safety systems with those already in place. In their minds the new shipments were no different from the cotton and other bulk commodities they handled. There was a general ignorance of the lethal potential of ammonium nitrate. The port authorities held the same views as the railway managers and ship's masters.

The fact that the nitrate had come from a U.S. Army ordnance factory should have raised questions about safety, especially as a 1941 army manual listed ammonium nitrate as a high explosive, with half the explosive power of TNT. In addition, a 1945 U.S. Department of Agriculture publication said that it would explode if given a strong impulse or held in a restricted space under conditions of rising heat and pressure. There were other conditions listed in different publications, such as the effect of fire or proximity to other substances, that added to the lethal potential of ammonium nitrate, and some of them directly related to the shipments at Texas City. No one seemed to know about the tragedy in Oppau, Germany, 26 years earlier.

In 1946 and the early part of 1947, 100,000 tons of ammonium nitrate passed through the port at Texas City for onward shipment to other countries. There had been no incidents during that time, reinforcing the assumption that it was not an explosive substance. The nitrate was manufactured in ordnance plants, where previously it formed part of the ingredients for bombs, and small quantities of clay were added to each shipment to prevent caking. Ammonium nitrate had the double advantage of being relatively inexpensive and richer in nitrogen than some other fertilizers. Following the end of World War II, principal destinations for the fertilizer were in Europe, where the United States sought to expand food production and speed up the war recovery. On April 16, 1947, the freighters *Grandcamp* and *High Flier* were in the harbor at Texas City loading cargo for shipment to France.

CAUSES

More than 2,000 tons of ammonium nitrate had been loaded onto the *Grandcamp* when a small fire was found in the hold, possibly due to a burning cigarette butt. Although smoking was officially prohibited on the docks, it was generally permitted at that time. Longshoremen who were working in the hold tried to put out the fire with fire extinguishers but failed. At that point the captain intervened. He felt that the use of water would damage other items in his cargo, so instead of dealing with the problem he closed the hatches and ventilators and turned on the steam system in an attempt to smother the fire. In addition, as a precautionary measure, he had cases of ammunition removed from a nearby hold. As the fire grew, the heat forced both longshoremen and crew to leave the ship. The ship's alarm was then switched on and the onshore fire department contacted.

Before help could arrive, hatch covers were blown off by the pressure buildup, smoke and flames shot upward, and moments later the entire ship disintegrated in one huge explosion. The Texas City auditorium was transformed into a first-aid center to cope with the thousands of casualties. Doctors, nurses, and ambulances were brought in from neighboring communities. Law enforcement officers also had to be called into service from other places to help establish order. But just as things began to come together, there was a second cataclysm.

Because of the general indifference to the dangers of ammonium nitrate, no one took much notice of the second freighter that was being loaded with the same cargo and was carrying 1,000 tons of it. The *High*

Flier had been torn away from its moorings by the force of the *Grandcamp*'s explosion and was stuck alongside other vessels, severely damaged. Before long the thick black fumes from the first explosion forced everyone to leave the *High Flier* and head for shore. Rescuers looking for survivors checked the freighter in the course of their search and noticed a fire in one of its holds. But there was so much anxiety over the first explosion that little thought was given to it.

Not until late that evening was any action taken to deal with the *High Flier*. By then flames were shooting high into the air. Before anything could be done to move the ship out to sea or douse the fire, the *High Flier* blew up in a second blast just like the first one.

CONSEQUENCES

The shock wave alone following the explosion of the *Grandcamp*, quite apart from the flying clusters of steel that accompanied it, did enormous damage. Two planes flying overhead were brought down, and everyone still on board the ship or near it was vaporized by the intense heat. Flying objects killed hundreds in the immediate vicinity of the pier, and lighter debris damaged buildings in the business district a mile away. Some of the flying pieces from the *Grandcamp* weighed several tons. A 15-foot wave swept up from the harbor by the explosion picked up a large steel barge and carried it ashore. There were quantities of cotton and other textiles in the ship's cargo hold, and these became fireballs raining down on shore and triggering fires in many locations.

At Galveston, 10 miles south of Texas City, people were thrown to the pavement by the force of the blast, and at some places twice as far away buildings swayed. A number of people rushed to the docks at Texas City to search for relatives and friends. Wounded people were everywhere, covered in black oil, stunned into passivity by their ordeal.

Casualties were much lighter following the explosion of the *High Flier*, not because the explosion was less powerful but because there already had been a general evacuation of the port area. This second blast did more damage to buildings that were already partly destroyed, oil tanks caught fire, and a shower of large pieces of steel, just as in the first explosion, caught anyone still near the piers. Additional fires were ignited.

The losses from the disaster were huge: 600 dead and thousands injured. An exact figure could not be obtained because many bodies were unrecognizable. No one in the city was unaffected by the events.

CLEANUP

By April 23 all of the fires had been extinguished and on May 16 the last of the bodies was removed from the rubble. Many weeks later federal and local authorities began to assess the damage and initiate new safety rules to ensure that a tragedy of this kind would never recur. Culpability was not acknowledged by anyone, yet errors due to neglect were everywhere. The whole infrastructure of the port area was wrong. Large oil tanks and storage facilities for other highly flammable products were located close to the piers without regard for the danger of a ship's catching fire. No procedures were in place to deal with emergencies.

The disaster brought changes in some of the processes involved in chemical manufacturing and were new regulations for the bagging, handling, and shipping of chemicals. Thousands of lawsuits were finally settled in 1956 at a total cost of $16 million. The bodies of 63 unidentified dead were buried together in a local cemetery. In the 1980s additional land was added to it and it was dedicated as a memorial park.

Liquid Natural Gas Explosion
Cleveland, Ohio
October 21, 1944

Liquefied natural gas (LNG), which is mainly methane gas, is a popular fuel for homes because it is odorless, colorless, noncorrosive, and nontoxic. It is used as a gas but stored in liquid form for compactness. To achieve the liquid state it has to be cooled to minus 250 degrees Fahrenheit and stored in well-insulated containers at that temperature. As a liquid its volume is reduced to 1/600 of the gaseous form and its weight is about half that of water. The commercial installation in Cleveland, Ohio, in 1941 was the second in the United States, and it served the community well for the next three years despite the fact that little was known at that time about LNG, particularly the dangers associated with its use.

The East Ohio Gas Company plant in Cleveland consisted of four steel storage tanks, one cylindrical in shape and three spherical. Each contained more than 200 million cubic feet of liquefied gas. A cooling tower stood beside the storage tanks to reduce the incoming gas to liquid form, and a steam generator was on hand to vaporize the liquid as needed and store it in a holding tank in its natural, gaseous form. This tank was sus-

Damage caused by the explosion at the East Ohio Gas Company's #2 Works, Liquefaction Storage Facility in Cleveland, Ohio, on October 20, 1944. The explosion, caused by a gas leak, sparked fires that burned 160 acres of businesses and neighborhoods in Cleveland. Over 100 people were killed. (© Ohio Historical Society)

pended in water and rose and fell in height as gas was either added or taken out to supply the city. There was one other important item in the complex, an old holding tank that received leaked gas from the main storage units and gradually released it into the atmosphere through a tall vent. Despite the high quality of the insulation on the storage tanks, leaks occurred as temperatures changed, and the old tank provided a safe means of dispersing it.

CAUSES

In 1944, a fifth storage tank was added to meet the growing demands of Cleveland consumers, but because it was wartime and steel could not be obtained for this purpose, an alloy was used. Before long the new tank began to leak, and small quantities of supercold gas escaped into the atmosphere. Emergency-repair workers were called in. There was no spare tank into which they could transfer the gas. All were filled to capacity because winter was approaching, so the only alternative was to repair the leak. This was done and the repair was rigorously tested, but in a short time it began to leak again. No one knows why the repair failed because all the men and the test results vanished in the explosion.

As the escaping gas entered the much warmer surrounding air, it formed a cloud over the east side of Cleveland, but authorities and residents were unaware of the extreme danger it posed. When mixed with air in the right quantities, LNG becomes a time bomb that will explode if touched by a spark. Fortunately neither liquefying nor vaporizing processes were being undertaken at that time. Achieving that mix of gas and air was inevitable as more and more gas escaped. Equally inevitable was the arrival of a spark somewhere in the city; even a shoe's striking a stone could cause one. A massive explosion and fire were triggered, and within moments neighboring homes exploded in flames.

CONSEQUENCES

The occupants of one home a block away heard the noise of the explosion soon enough to run away. Others joined them, and as they looked back they could see that flames had risen a half mile into the air. A second explosion followed as a second tank succumbed to the heat and exploded. Then gas lines ignited all over the eastern part of the city. Every building for 50 blocks was destroyed, and manhole covers were flung into the air like toys, adding to the damage as they returned to earth. The firestorm from the explosions destroyed more than 150 homes and as many offices and left 1,500 people homeless. One hundred thirty people died and more than 200 were injured.

CLEANUP

The cost to the city for reparations amounted to $15 million. Cleveland's tragedy was little known at the time because of all the action of World War II. Even the news of the dead and injured did not seem all that important when large numbers of deaths were being reported from the war fronts. The biggest entry in one newspaper was a short statement on the loss of hundreds of cars in one parking lot. As a result of the event, no

Relatives and victims of an explosion of liquid petroleum gas, which destroyed the slums known as "San Juanico" and killed 500 people, attend a religious ceremony to mark the anniversary of the tragedy, in Mexico City. (© Notimex/NewsCom)

additional LNG installations were built for 20 years and extensive research was undertaken to ensure that an explosion of this magnitude would never occur again. New regulations were introduced governing the choice of materials for storage tanks, procedures for transportation of LNG, and the location of storage tanks at a distance from residential sites. Other rules prescribed ways of testing tanks and installing safety valves on pipelines.

Today there is widespread confidence in the safety of LNG plants; and in fact, Cleveland was the first city to feature a new use for LNG. The Greater Cleveland Regional Transit Authority adopted natural gas in 1995 as its alternative fuel of choice for buses. It built the biggest natural-gas fueling station in the country, and Cleveland has served as a model for the rest of the nation. Throughout the 1990s, gas-powered buses were still being added to Cleveland's streets every year, and air pollution was reduced as each new bus joined the fleet.

Liquid Petroleum Gas Explosion
Mexico City, Mexico
November 19, 1984

In the 1940s the Mexican government, through its state oil company Pemex, decided to store liquefied petroleum gas (LPG), or propane, in selected centers around the country and distribute it to consumers by truck. San Juan Ixhuatepec, eight miles north of Mexico City, was one of these storage centers. Twenty-four-gallon drums were loaded daily and trucked to the surrounding communities. It was a good arrangement from the consumers' point of view because the gas reached most people quickly and most Mexicans cook on propane stoves.

The climate of an area is critical for the safe storage of LPG. At 44 degrees Fahrenheit below freezing, the gas becomes stable, and it will remain so in any temperature regime provided the insulation is excellent. Mexico City is in a warm climatic zone and temperatures there rise very high at times. To allow for the impact of temperature change on the storage tanks at Ixhuatepec, they were designed to cope with temperatures much higher than normal, and Pemex claimed that the insulation was completely satisfactory. A rise in internal temperature means an increase in internal pressure because some liquid gas always vaporizes under these circumstances.

CAUSES

On the night of November 18–19, 1984, a worker reported that the pilot flame at the San Juan Ixhuatepec plant had gone out, something that should never have happened because the pilot serves as a warning signal. If it went out, that meant that significant amounts of gas were escaping. Near the pilot flame a loud hissing noise was heard. Several other workers noticed a strong smell of gas over the whole area for several hours. Evidently a substantial amount of gas had escaped, yet no one had reported the problem. Early the following day, just before 6 A.M., an explosion was triggered by ignition of the escaping gas. Whether the quality of insulation was inadequate or there was some other weakness in the storage tanks, the pressure rose high enough to force open a relief valve and allow a cloud of gas to escape. A series of fires followed, and then came a much more powerful explosion as two very large cylinders were heated by the fires and blew up. More explosions and fires followed as other tanks caught fire.

CONSEQUENCES

Fireballs erupted all around the plant whenever another cylinder or pipe blew up. Masses of fragmented metal rained down. One six-foot section of piping was flung two miles away. Another section landed on a house half a mile from the plant, killing a number of people. In one nearby area 200 houses were totally demolished by a fireball. Most of the occupants died in their sleep. For several hours there was no organized evacuation; people just fled to the hills to get away from the fires. Some badly hurt people were left to die on the street. One of the explosions was so powerful that it was measured at 0.5 on the Richter scale in the seismograph station at the University of Mexico about eight miles away.

The whole Pemex complex was destroyed along with all the workers' homes. The 30 acres on which the plant and homes stood had become a wasteland.

CLEANUP

The main supply gas line was closed down immediately and 80,000 containers of gas were burned off to prevent them from exploding. Firefighters battled the flames all day, finally putting out the fires by nightfall. Emergency aid stations were set up to cope with the thousands of injured and homeless people. The International Red Cross and other aid agencies appealed for blood and medical supplies through the local media. At a nearby sports complex arrangements were quickly made to accommodate and feed more than a thousand families.

The first priority was disease avoidance. The bodies of the dead and sewers that had broken open were attractions for rodents and flies, so extensive fumigation was carried out. Then came the task of burying the dead. The true count was not known. Many of the workers' identities had not been recorded. The official figure stood at 500. Five thousand more were injured and tens of thousands lost their homes. A day after the tragedy, coffins containing the remains of 300 victims were placed in two huge holes, each 500 feet wide, in the hillside of a cemetery close to Ixhuatepec. Health workers walking around the area to give inoculations against typhoid and tetanus often were met with extreme anger because of the intense suffering. Many of the victims knew of earlier Pemex accidents in the same area, and they felt that this one could have been prevented.

Their feelings were confirmed when Pemex tried to blame the accident on something that happened at a neighboring plant. For more than nine months Pemex insisted that it was not responsible for the tragedy. Then came a report from the Netherlands Organization for Applied Scientific Research, which tracks accidents of this kind all over the world. It provided a detailed analysis of the event. The claim by Pemex that the disaster started at a neighboring plant was rejected. The Netherlands Organization confirmed that a leak from a pipe in the installation at San Juan Ixhuatepec was the cause. Subsequently Mexico's attorney general placed the blame on Pemex, accusing it of mismanagement and disregard of safety precautions and ordering it to pay damages as the courts determined them.

About 150,000 people had lived close to the San Juan Ixhuatepec plant, in violation of all the rules governing industrial safety. There should have been no homes near such a large plant. In Mexico City, however, housing costs are prohibitive for Pemex workers. They take the risk of living next door to the huge LPG cylinders, and company officials do not interfere. The workers were the ones who were killed or injured. Even in the aftermath of this terrible tragedy, when what remained of the neighboring homes had been cleared away, new homes began to appear where they were before.

A body is taken away from the remains of the Nypro Chemical Plant at Flixborough after a massive explosion ripped through the plant. (© Hutton/Archive)

Benzene Explosion
Flixborough, England
June 1, 1974

In 1974 the nylon plant at Flixborough, England, was the only one in Great Britain producing caprolactam, a key component in the manufacture of nylon. Nylon was in high demand at the time, and world supplies were barely able to keep up with demand. It was used in the manufacture of hosiery, carpet backing, seat belts, and car tires. In various locations throughout Britain, 30,000 workers were employed in these nylon-based industries, so there would be serious economic consequences if a breakdown occurred at Flixborough.

Prior to the 1960s no British firm manufactured nylon because a U.S. manufacturer controlled the only known process for making it and had patented the process under the name Nylon 66. A Dutch company succeeded in finding a new way of manufacturing a synthetic that was almost identical to nylon, and it arranged a partnership with the British Coal Board for its production. By the late 1960s the Flixborough factory was supplying 20,000 tons of caprolactam every year to Courtaulds, Britain's biggest manufacturer of synthetic textiles.

CAUSES

Over time, the manufacturing process was sped up through improved methods until by early 1974 annual output was 70,000 tons. It was then that the factory managers at Flixborough decided to switch to a time-saving shortcut that involved the use of benzene. The Dutch company had other factories in North and South America, all producing the critical caprolactam, but only the British plant switched to benzene for a quicker and cheaper method of production. Benzene is toxic, flammable, and it readily explodes. It needs very special care and precision handling because the production process involves very high pressures. The risks associated with its use are therefore much greater.

A few days before June 1, 1974, a pipe through which benzene was carried at high pressure developed a leak and had to be removed for repair. A temporary pipe was installed at once in its place in order to maintain production levels. There was no professional mechanical engineer on site at the time to supervise this work. The men who were asked to find and install the temporary pipe had plenty of practical experience and technical ability. They found piping of the correct dimensions and had the plant back in operation in a few days. They did not know the nature of the processes involved in the production of caprolactam and so they made no provision for them. For example, pipes that have to cope with very high temperatures and equally high pressures, especially 20-inch diameter ones as in this case, demand expert design and the right material.

On Saturday afternoon, June 1, when most workers were off duty, a technician noticed small flames moving along a pipe. Benzene was leaking. He was well aware of the danger of even the smallest spark near benzene and immediately shouted to those around him, urging them to run as fast as possible.

CONSEQUENCES

Eight men were able to escape before a massive explosion erupted, totally destroying the plant. Another man had a different kind of escape. He was flung 30 yards by the force of the explosion and then wandered, dazed, for some time. Fortunately, the number of workers on site was small when the explosion happened: only 70 of the normal complement of 550. It was later discovered that the temporary pipe had failed to cope with the pressures to which it was subjected.

One observer six miles away thought the blast was an atomic bomb. The force of the explosion was equivalent to detonating 16 tons of TNT. The small community adjacent to the plant was in total shock. Everyone thought that chemical factories of this size and importance would be so well protected that serious accidents could never happen. These neighbors had little to do with the plant. None of them worked there, and they had never been informed about the nature of the caprolactam production and the risks associated with it. It was one of the most serious accidents in the history of Britain's chemical industry. Twenty-nine people were killed and 100 others injured. The plant was totally destroyed, while 800 homes and 160 stores were damaged. Reconstruction costs and compensation payments amounted to $100 million.

CLEANUP

One of the concerns raised after the accident was the folly of building large chemical factories beside residential communities, and new laws were introduced to prevent a repeat of the Flixborough-type site. The risk factor had been given too little attention by the British government. Nine months before this accident the government's chief inspector of factories was concerned about the jumble of old regulations that dealt with industrial hazards. He proposed a new bill, Health and Safety at Work, to replace all the old laws but did not get it through Parliament before the Flixborough tragedy. Many people wondered why it took so long to think about a new approach like this.

Largely as a result of the Flixborough explosion, European nations in 1982 created a directive, a wide-ranging document that sought to prevent the recurrence of chemical accidents. The countries involved were particularly concerned about the dangers that can arise when original factory designs are changed, as was the case at Flixborough when benzene replaced the materials used at other similar factories. This directive was later amended several times to strengthen its requirements, in response to additional damaging chemical spills in several countries.

SELECTED READINGS

Benson, Ragnar. *The Greatest Explosions in History*. New York: Carol Publishing Group, 1991.

Butler, Paul. "Lessons to Be Learned from the Flixborough Enquiry." *Engineer* 11 (11 December 1975).

Pearce, Fred. "After Bhopal, Who Remembered Ixhuatepec?" *New Scientist* 107, no. 1465 (18 July 1985).

Stephens, Hugh W. *The Texas City Disaster, 1947*. Austin: University of Texas Press, 1997.

5

Nuclear Energy Accidents

Three Mile Island
Pennsylvania
March 28, 1979

There are dozens of nuclear power plants operating in the United States, providing electrical power to more than half of the nation's states. An accident in any one of them sends alarms across the country. So when something went wrong with the installation at Three Mile Island (TMI) in Pennsylvania in March 1979, it cast a shadow on every nuclear power plant in the rest of the country. The companies that operate these plants know that their power generators have backup systems to cope with accidents, and they also know that if these systems fail there could be a serious tragedy. The basic rule in operating nuclear generators is to have duplicate safety systems for each of the important operations so that the failure of one will not cause irreparable damage before repairs can be completed. Such duplicate systems were installed in TMI, but because nuclear power generators were still relatively new at the time not every eventuality was provided for.

Unit number two at the TMI nuclear power plant was owned by the Consolidated Edison Company and had been built on an island in the Susquehanna River about 10 miles south of Harrisburg, Pennsylvania. It became fully operational late in December 1978. The heart of the power plant is the reactor (see figure 5.1), which heats water by nuclear fission.

Aerial view of Three Mile Island. (© National Archives)

The hot water in turn produces steam that drives a turbine. The turbine then drives a generator that produces electricity. Cylindrical pellets of uranium fuel are stacked inside the thousands of fuel rods that constitute the nuclear reactor, and these rods are then encased in a tank of cooling water to moderate the temperature. The constant supply of cooling water is vital, and hence there is always an emergency supply of water available.

CAUSES

At about 4 A.M. on March 28, 1979, the pumps providing cooling water to the nuclear reactor of unit two at TMI failed, and alarm bells started

President Jimmy Carter leaving Three Mile Island four days after the accident. (© National Archives)

ringing all over the control room. The operators immediately switched on the emergency system but—unknown to them because they could not see the warning lights on the pumps—the emergency pumps also failed. Within a couple of minutes both the steam turbine and the electric generator automatically stopped. It was discovered later that the water pipes had been undergoing a routine cleaning over the preceding few days, and in the process one of the pipes became blocked. While trying to clear it, an operator accidentally switched off the main flow of water to the emergency system.

The flow of steam removes some of the intense heat generated in the reactor's coolant water. This removal of heat is one of the laws of physics; namely, that some heat is lost when water turns to vapor. The reverse happens when water vapor turns back to water; in that case, heat is added. The absence of steam for the turbine meant that the temperature of the coolant water in unit two's reactor began to rise.

When the temperature reached a critical level, a pressure-relief valve automatically opened, allowing steam and water to flow out of the reactor through a drainpipe to a tank on the floor of the main structure. At this point, less than 10 seconds after the first failure, because no fresh supply of

Figure 5.1
The Heart of a Nuclear Power Plant (Paul Giesbrecht)

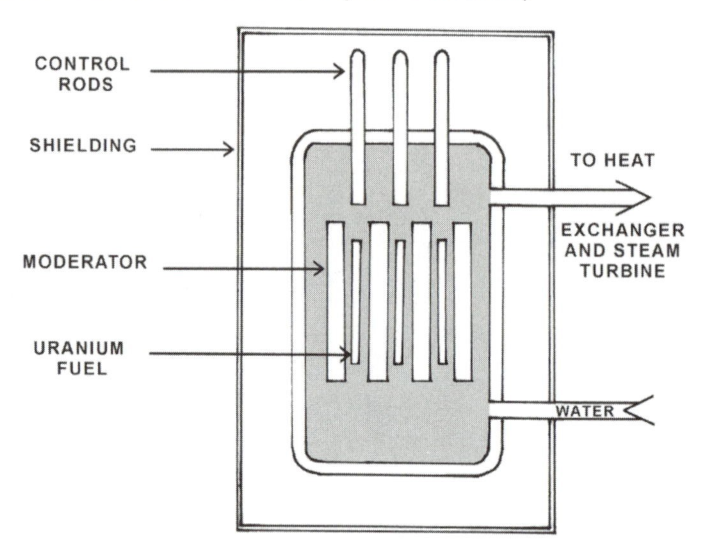

water was forthcoming, the reactor took the next preplanned step and shut down. Shutdown is accomplished by letting the control rods drop from above into the reactor core, stopping nuclear fusion. These rods can be raised or lowered by remote control, and they can be partially raised when the power plant is running at less than full capacity. The operators heard the loud, characteristic noises that confirmed the shutdown. Within a few seconds, had things gone according to plan, emergency water supplies would have begun to flow, and the plant would be ready to start up again. But a third unexpected problem now arose: the pressure-relief valve stuck in the open position so that thousands of gallons of coolant water continued to drain away from the reactor.

The operators, not knowing that the level of coolant water had dropped dangerously low, exposing the reactor core and risking a meltdown, made no effort to restore water to the core. Within a few minutes, as temperatures within the core continued to rise, the remaining water turned into steam and the fuel rods in the reactor core overheated and began to disintegrate. As they did, some of the exposed material in the rods interacted with steam to produce hydrogen. As this mixture moved down the drainpipe to the tank, an explosion occurred, releasing radioactive gas into the surrounding community.

CONSEQUENCES

Three hours after the initial problem, the plant manager arrived and a state of emergency was declared. Preliminary warnings were circulated to people in the immediate vicinity as radioactive material and water continued to escape from the plant.

A large number of schools in the region were closed down. First, students inside every school were assembled in one place in the building and all windows were closed. The students were then taken by sealed buses to other schools 12 miles away from the nuclear power plant. All other places near the reactor were also evacuated, and extensive tests were made of the levels of radiation at various distances from the source. A complete meltdown did not take place, that is, nuclear fuel did not melt through the floor beneath the containment or through the steel reactor vessel. However, a substantial amount of fuel did melt. Radioactivity in the reactor coolant increased dramatically. Radioactive gas, mainly xenon, spread through small leaks. It reached all parts of the plant and went out into the surrounding environment.

The general sense of alarm in the community continued for almost two weeks before the Nuclear Regulatory Commission was able to declare that the crisis was over and it was safe for everyone to return to their homes and places of work.

CLEANUP

Detailed studies of the area surrounding TMI were continued by a number of government agencies and several independent ones. The general conclusion was that the average radiation exposure to about 2 million people in the area was one millirem. A rem is the standard unit for measuring the radiation absorbed by a human, and a millirem is one-thousandth of one rem. One millirem is equivalent to one-sixth the total radiation exposure of a full set of chest x-rays. The natural radiation level in this part of Pennsylvania is a little more than 100 millirems per year. Clearly, the average amount of radiation exposure as a result of the accident was trivial, and the maximum that anyone was likely to experience was the same as everyone experienced in this area.

In the months that followed, although questions were raised about possible adverse effects of radiation on human, animal, and plant life in the area, none could be directly linked to the accident. In the process of coming to this conclusion, thousands of environmental samples of air, water, milk, vegetation, soil, and foodstuffs were collected and monitored.

The TMI event was what one expert called a "common-mode accident"—an event that bypasses all the backup systems either because of its rarity or because an operator interferes with safety systems by disconnecting them. The fundamental weakness that gave rise to the common-mode accident at TMI was the incompetence of the plant operators. These maintenance people, who looked after the installation day and night, were well trained for keeping everything in good order, for conducting routine adjustments and repairs on equipment, and for reporting abnormalities. They had spent a year in a training course, and some of them had previous experience on nuclear-powered submarines. They were not qualified nuclear engineers, and they had no training for coping with complex emergencies—nor were they supposed to. The automatic systems were designed for these emergencies but, as it turned out, they could not anticipate every eventuality.

Causes of the TMI accident continue to be debated to this day. The main factors appear to have been a combination of personnel error, design deficiencies, and component failures. There is no doubt that the accident permanently changed the nuclear industry. Public fear and distrust increased. It was the most serious accident in U.S. commercial nuclear-power-plant operating history, even though it led to no deaths or injuries to either plant workers or members of the nearby community. One outcome was that no nuclear power plants were built during the rest of the twentieth century. More recently, with increasing demands for national self-sufficiency in energy, pressure began to build for beginning construction again.

President Carter appointed a commission to study the causes of the accident and make recommendations. Its report was a scathing condemnation of the work of the Nuclear Regulatory Commission (NRC), and the commission's members recommended fundamental changes in the NRC's organization, procedures, practices, and attitudes. At the heart of the criticism was the feeling that the NRC was mostly concerned with equipment problems, yet all the errors uncovered, including those at TMI, were the fault of people. Furthermore, lessons from previous accidents caused by human error had not been passed on to the operators at TMI. One particular case from thirteen months earlier was so similar in operational details to the situation at the TMI plant that it was essential that TMI operators know about it, yet nothing was done to inform them.

The commission's report brought about sweeping changes in emergency-response planning, reactor operator training, and many other areas of nuclear power plant operations. Reactor operator training was highest on the list of reforms. All electric utilities expanded their training for personnel who work at and support nuclear power plant operations. In addition, modern information technology was applied in new ways to the day-to-day operations at plants. For example, at TMI's control room, a hundred alarms went off within the first few minutes of the accident, but there was no system in place to prioritize these signals in order of importance.

The National Nuclear Academy was instituted to accredit training programs. Utilities purchased simulators for the training of personnel who work in control rooms. Training reforms centered on protecting a plant's cooling capacity, whatever the triggering problem might be. In the 1979 accident, operators turned to a book of procedures to pick those that seemed to fit the event. In the new training, operators are taken through a set of yes/no questions to ensure first that the reactor's fuel core remains covered. Then they determine the specific malfunction. This is known as a symptom-based approach for responding to plant events. Underlying it is a style of training that gives operators a foundation for understanding both theoretical and practical aspects of nuclear installations.

The cleanup of the damaged nuclear reactor took nearly 12 years and cost almost $1 billion. The work was challenging technically and radiologically. Plant surfaces as well as the water used in the cleanup had to be decontaminated. One hundred tons of damaged uranium fuel had to be removed from the reactor vessel without harming the workers involved. Nuclear waste material was sent to Richland, Washington, for storage. After the cleanup, the reactor was placed on long-term monitored storage. It was kept completely free from its neighboring unit, number one, which, although unaffected by what had happened, was shut down at the time of the accident. Unit number one was restarted in 1985 and has been working efficiently and safely ever since.

Chernobyl
Ukraine, Soviet Union
April 26, 1986

The Chernobyl nuclear power plant near Kiev in Ukraine, formerly a state of the Soviet Union, was the pride of the country's nuclear power program. It was the biggest installation of its kind, with four generators, each producing 1,000 megawatts of electricity, and two more of the same kind were under construction. Work on the first unit began in 1971 and by early 1986, with four units operating, the Soviet government was about to launch a five-year program in which several new Chernobyl-style plants would be built. The decision to go ahead with this was made in

The Chernobyl nuclear power plant a few days after the explosion. In front of the chimney is the destroyed fourth reactor; behind the chimney and close to the fourth reactor is the third reactor, which was stopped on December 6, 2000. (AP/Wide World Photos)

February 1986, and the Communist Party Congress that June was to approve it. Between those two months came the explosion of Chernobyl's reactor number four.

CAUSES

At the time of the accident, reactor number four was undergoing routine maintenance that required a shutdown. The manager of the plant decided to conduct an experiment during this time, a test that could only be conducted during a shutdown. He wanted to know how to deal with a reactor problem if power were unavailable from the main electrical grid. Even when a reactor is shut down it requires power to maintain the cooling circuit. The plant manager's experiment was to use diesel generators in such an emergency, and he hoped that the 50 seconds or so that these engines required to reach full speed would be sufficient to maintain the safety of the power plant. All the emergency safety systems were shut down for the experiment. This included cutting off the emergency cooling system. Even a warning printout showing that the reactor was in danger of overheating and should be shut down immediately was ignored.

The initial events in the accident were similar to those experienced at other nuclear facilities. The reactor lost its cooling water and the nuclear fuel elements began to heat up. Action can normally be taken at this stage to bring in emergency supplies of water so time is gained to locate the trouble and fix it. In the case of the Chernobyl accident, nothing was done to counter the buildup of heat. Worse still, all the warning signals were ignored. The reactor did not have a protective building around it, like U.S. installations, to prevent the escape of radioactive gases. Shortly after 1 A.M. on April 26 the amount of heat in the reactor reached a critical level, yet the operators who noted this did nothing. Less than two minutes later the graphite in the reactor core caught fire. A burst of radiation escaped into the surrounding area, followed shortly afterward by solid radioactive material.

CONSEQUENCES

The accident was so powerful that a plume of radioactive material and radiation shot upward into the sky and was carried northward by the wind across Poland and Scandinavia. A day later technicians at a nuclear facility in Sweden picked up high levels of radioactivity. It was the remnants of a much thicker cloud that caused the most damage in the area around the Chernobyl plant and the second-greatest amount of damage as it passed over the state of Belarus. Nothing was said about it then, nor was much revealed for days afterward. Characteristic Soviet secrecy surrounded the event. Some reports were lost or destroyed. The Soviet Union was anxious to maintain a positive approach to nuclear sources of energy because of its plans to expand them. They did not want to arouse public fears. All this secrecy was maintained in the face of accurate information available to Soviet authorities telling them exactly how much radiation had reached any part of eastern or northern Europe. A previous agreement with the United States required that each side have the ability and the equipment to determine this. In full knowledge of the terrible consequences of exposure to radiation, the Soviet Union refused to make public what it knew until compelled by external evidence.

The long-term effects of radiation have yet to be understood. They modify the genetic structures of animals, food, and humans. The Chernobyl accident led to huge health problems in the Ukraine and elsewhere. Plant personnel, fire fighters, medical staff, and cleanup workers suffered the most. Reports from Belarus have indicated a 50 percent drop in birthrates and a steady rise in miscarriages and birth defects. When the countries of Western Europe refused to accept agricultural produce from Eastern Europe, the Soviets claimed their refusal was based on propaganda. But researchers were able to counter the Soviet claim that no other country was affected by radiation from the accident. Within a year the evidence of widespread devastation was overwhelming, and the Soviet Union was forced to accept it. Scandinavia, Great Britain, West Germany, Italy, and all the countries in between had been hit with damaging radiation and a very large area was contaminated.

Sweden was one of the hardest hit, with much of the damage being carried in rain. An area of 4,040 square miles 175 miles north of Stockholm was so badly contaminated that all the grass had to be harvested and burned. Thousands of gallons of milk from this area was discarded daily for some time. In Sweden's far north, where the Sami (formerly Lapp) people live, levels of radiation were 12 times the permissible limit. These people were dependent on reindeer for almost all their needs. The Swedish government helped pay the costs of destroying about 50,000 reindeer. To make matters worse, the vegetation of northern latitudes has a very slow rate of decay for radiation contamination. In 1987 it was estimated that much of the vegetation and soils in the region where the Sami live will remain contaminated until 2030.

Thirty-one people died in the initial explosion and fire, and millions were exposed to radioactivity. The final toll of casualties will never be fully known, but it will far exceed the five million initially identified as victims of radiation.

CLEANUP

The staff at the Chernobyl nuclear power plant attempted to assess the extent of the damage to unit four and limit the spread of fire to the other reactor units. By doing so, many of these people averted what could have been a far greater catastrophe, but they lost their lives as a result of lethal doses of radiation. Firefighters poured water into the burning reactor. Over a period of two weeks, the Soviet air force dropped more than 10,000 tons of material into the reactor core to try and smother the fire. The pilots who flew these missions died from the massive radiation doses they received, and a dozen huge helicopters became so radioactive that they were dumped in open pits along with trucks, cars, and other contaminated objects around Chernobyl.

In a reactor core, when heat rises sufficiently high, hydrogen explosions occur. In unit four, the explosions hurled burning lumps of graphite and reactor fuel into the air; those then landed on neighboring buildings and set them on fire. Most of the buildings had tar roofs, and they fueled the fires. Once the fires were extinguished, there was the question of what to do with all the radioactive debris that had escaped from the reactor core.

It was decided to gather as much as possible and push it back into the reactor. This dangerous task was at first undertaken by robots, but they were unable to cope with the terrain. When the robots kept getting stuck, a fateful decision to use humans was made.

Choosing to use people for this task was tantamount to killing them. Everyone involved with nuclear installations knew the deadly effects of radiation by this time. The men, mostly from the army, were only able to work for a maximum of one minute even with heavy lead protective clothing on. Radiation levels were dangerously high. The one minute's work proved to be too much for some. They later succumbed to serious illness. A massive concrete container was built to encircle the damaged reactor and form a roof on top. It took seven months to complete, stood 200 feet in height, and stretched for 200 feet along each side of the reactor building. Construction was speedy, and the quality of work correspondingly poor. Everyone wanted to minimize the workers' exposure to radiation. The containment barrier was intended to last for 30 years, but within a decade cracks and weaknesses appeared and repairs were needed. In addition the concrete was gradually weakened by irradiation from within and by tension from the huge temperature differences between inside and outside.

Scattered around the Chernobyl nuclear power plant were hundreds of dumps of radioactive waste. These usually consisted of open pits with linings of clay, containing anything from soil, timber, and vehicles to domestic items such as refrigerators and clothing. Some of the pits contained the remnants of the forest that had surrounded the power station and absorbed so much radiation that the trees had to be destroyed and treated as radioactive waste. These pits became an environmental hazard because they posed a threat to the area's main water table. The water table is linked to the Dnieper River, which supplied the water needs of 35 million people, including the residents of Kiev.

Police and military personnel guarded the area around Chernobyl, ensuring that no one ventured close to lethal radiation. In spite of that, a few people, mostly elderly, were allowed back into their homes. They insisted that because they could not see, taste, smell, or touch the deadly radiation it didn't bother them. While they were not concerned, future generations must be.

In December 2000 the last of Chernobyl's four reactors was shut down permanently, the final act in a financial deal struck between western nations and Ukraine. The nuclear power plant and an area extending outward for 19 miles in all directions became a wasteland. The concrete barrier erected around the plant at the time of the accident still needs repair from time to time.

Tokaimura
Tokai, Japan
September 30, 1999

Japan has experienced a series of nuclear accidents in recent years. Since 1995 there have been seven mishaps, and fears are mounting over the dangers inherent in this source of energy. Japan is the only country to have experienced the horrors of a nuclear bomb so people have a very understandable dislike of nuclear technology. One in the string of accidents occurred at the Turugu nuclear power station in July 1999, when there was a leak of cooling water. Unfortunately, the Japan Atomic Power Company (JCO), which is responsible for the station, suggested that the accident was a minor one. It was later confirmed that contamination levels were 11,000 times higher than the safe limit, so the accident affected public confidence in the system. Two months after the Turugu leak, on September 30, Japan's worst nuclear accident occurred at its Tokaimura power station, 87 miles to the north of Tokyo.

CAUSES

Three workers at the Tokaimura power station were close to finishing a process of purifying uranium oxide, getting it ready for use as fuel rods in the reactor. It was a job done only occasionally, and for two of the work-

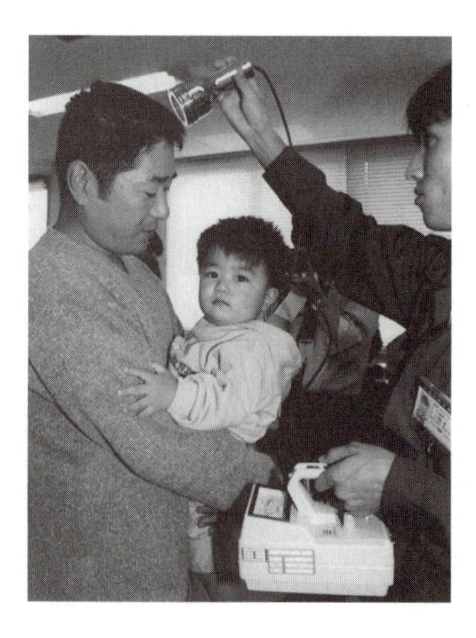

A family having its radiation level checked by a geiger counter at a community center in Tokaimura, in northern Japan, Friday, October 1, 1999. Under advisory to stay indoors one day after Japan's worst accident at a nuclear plant, this town was buried under an eerie silence Friday as worried awaited proof that it was safe to once again venture outdoors. (AP/Wide World Photos)

ers it was their first experience with the process. They had never previously handled uranium oxide. The third man had not worked at this process for three years. This final phase involved dissolving uranium oxide in nitric acid, and the three men were running late in their schedule. Uranium oxide can only be handled safely if the quantities worked on at any one time amount to less than seven pounds. Strict instructions had been given to keep within that limit. At eight or more pounds a critical mass is formed and a nuclear reaction can take place, with lethal radiation being emitted. Safety-designed containers were specially chosen for this work. They were small and narrow with a high surface-to-volume ratio. This ensured that neutrons would escape in sufficient quantities to prevent a chain reaction from occurring. Neutrons determine whether a particular form of uranium is stable or radioactive.

The small size of these containers, however, meant that the process took a lot of time. To save money, the company operating the plant decided—secretly, and in violation of standard procedures—to use stainless steel buckets instead of the special containers. This would enable them to halve the time needed for the work. The same illegal procedure was repeated when the processed uranium was to be moved for a time to another safety-designed container. On this occasion, the larger tank they selected was encased in a cooling jacket that reflected neutrons back. No one seemed to be aware of this additional danger. It increased the risk of a nuclear chain reaction and provided support for its continuance if one happened.

The safety of the three workers and all others in the vicinity had already been compromised by changing official safety regulations. At this point, in order to speed up the work, the men decided to increase the amount of uranium they placed in the container to more than 40 pounds, thus guaranteeing a chain reaction and a burst of lethal radiation. An intense emission of neutrons and gamma rays shot outward, and the men saw the telltale bright blue light that indicated the presence of lethal radiation. Almost immediately, two of the three workers fell ill with radiation poisoning. The third man, who was standing farther from the container, was less severely affected. Alarms were sounded and evacuation began.

CONSEQUENCES

Tokai, in which this plant was situated, has 15 other nuclear installations, all of them less than a few miles from the village center. All of this area lies within the six-mile emergency evacuation zone established for the Tokaimura nuclear power station, but the village was poorly prepared for any emergency. There were no neutron detectors to alert people to the fact that the chain reaction was still continuing. Without this data, most of the officials in the surrounding area assumed that the chain reaction had come to an end because this was the usual outcome in an accident of this kind. These officials did not know that the excessive amount of uranium that had been placed in the container both triggered and sustained the radiation.

Three paramedics responded to the alarm within 15 minutes, but they were unaware of the nature of the accident. They had no protective gloves or instruments to measure radiation, and they had to be told to move the injured away from the site as quickly as possible before trying to help them. The paramedics received doses of radiation as high as 1.3 rems while others were affected by quantities ranging from 1.5 to 4.7 rems. Every one of these measurements is thousands of times above the safe limit, and exposure to these levels means that those affected will experience either crippling diseases or death. There were no facilities in Tokai to treat the injured, so they were transferred to the National Institute of Radiological Sciences, 60 miles away. The two workers who were closest to the source of radiation were transferred a second time, to Tokyo's University Hospital. One died there within three months in spite of heroic efforts to save his life, including transplanting stem cells. The other died the following year.

CLEANUP

The various agencies in Tokai responsible for dealing with the emergency seemed to be in great confusion. To begin with, nothing was done for about three hours after the alarms sounded at about 11:00 A.M., then officials told everyone to stay indoors, thinking that would be sufficient to protect them from radiation. Fortunately, the mayor of Tokai decided by three in the afternoon to order everyone within 1,000 feet of the plant to move out, but the evacuation was not completed before eight in the evening, nine hours after the first alarm sounded. The problem was far from being resolved even then. Testing equipment, which by now had arrived at the plant, revealed that radiation was dangerous for far greater distances than 1,000 feet. Because of all the delays and errors, more than 700 people in and near the plant were exposed to doses of radiation far above the safe level before the chain reaction was finally stopped. The International Atomic Energy Agency regarded the accident as the world's third worst after Chernobyl and Three Mile Island.

Police were called in when it was discovered that JCO had bypassed safety regulations by not using the proper containers for uranium. It became obvious in the course of the police investigations that JCO had been using illegal procedures for more than three years, and this reflected badly on the regulatory authorities, which were required to inspect all aspects of the company's operation annually. The authorities concerned stated that they were not required by law to inspect occasional operations like this one. Within three months of the accident, the government of Japan passed two new laws to correct the worst deficiencies of the inspection system.

The authorities in Japan always point to the inherent safety of the technology in the nuclear industry. They insist that there are only miniscule risks of a technical failure, perhaps one chance in many millions. The most significant nuclear disasters of the past 30 years have been caused by human error, however, not technical failure. Furthermore, Tokaimura had two accidents in 1997, both caused by human failure. Two fires in barrels of radioactive waste caused one of them, and rainwater entering a storage pit for radioactive waste led to the other, endangering water supplies.

Alongside the accidents already experienced and the people's fears, there is the risk of earthquakes. They occur in Japan every day at some magnitude and at times at very high magnitudes. Some of Japan's nuclear stations may be dangerously close to an active fault, and that is totally unacceptable in the nuclear industry. Not every nuclear installation was given the detailed amount of geological study needed when it was built, and some experts are now saying that these sites should be revisited. Public fears of nuclear technology, earthquake risks, and problems over the past five years add up to a mounting opposition to nuclear energy. Consequently, the government is taking a fresh look at the future of nuclear power generation.

SELECTED READINGS

Ford, Daniel, F. *Three Mile Island: Thirty Minutes to Meltdown*. New York: Viking Press, 1982.

Gray, Mike, and Ira Rosen. *The Warning: Accident at Three Mile Island*. New York: W.W. Norton, 1982.

Lewis, Elmer E. *Nuclear Power Reactor Safety*. New York: John Wiley and Sons, 1977.

Medvedev, Zores. *The Legacy of Chernobyl*. New York: W.W. Norton, 1990.

6

Oil Spills

Amoco Cadiz
Coastal Brittany, France
March 17, 1978

The supertanker *Amoco Cadiz* sailed from the Persian Gulf early in February 1978 with a cargo of 223,000 tons of crude oil. The five-year-old ship was fully equipped with all the latest navigational aids. Its destination was Rotterdam, in the Netherlands, and the voyage involved sailing around South Africa, then northward through the Bay of Biscay into the English Channel. Traveling at 15 knots (slightly more than 15 miles per hour), it reached the northwest coast of France after five weeks. Passing through the Bay of Biscay, it encountered rough seas and storm-force winds, not uncommon in the area but unusually strong at that time.

The vessel was low in the water because of the full cargo it carried. As waves washed over the decks and the ship rose up and then plunged down in the water, the captain was concerned about the stress being put on the tanker. The vertical movements were matched by a rolling movement, and to counter this the helmsman switched from automatic steering to manual. He then swung the rudder rapidly, first to one side and then to the other, hoping that by doing so he would reduce the amount of rocking and rolling. At the same time, he turned the ship into the oncoming waves and reduced speed. All of this activity was necessary to keep the ship on course as it approached the entrance to the English Channel.

The fully laden 233,000-ton Liberian supertanker *Amoco Cadiz*, as it sinks into the ocean near Portsall, in Brittany, France, on March 17, 1978. A full pollution alert was ordered and the French navy evacuated 44 crewmen. (AP/Wide World Photos)

CAUSES

On the morning of March 16, the *Amoco Cadiz* was about 12 miles west of the French port of Portsall when the helmsman suddenly called out that something was wrong with the steering. It soon became clear that the rudder had completely failed and the ship was out of control, perhaps as a result of the great pressure experienced in the storm. The captain immediately ordered the engines stopped, hoisted two black balls to show that his ship was no longer under control, and then broadcast a distress signal to alert other vessels and get help. It took 20 minutes for the ship to stop. A coastal radio station reported the presence of a German tug only 15 miles away, and by 11:20 A.M. it was on its way to the *Amoco Cadiz*, expecting to reach it within an hour. Meanwhile, the supertanker's captain had contacted Amoco's head office in Chicago by telephone. The *Amoco Cadiz*, more than seven miles off the French shore, was slowly drifting toward it.

The German tug arrived at 12:20 P.M. and immediately began to move into position to get a tow cable on the *Amoco Cadiz*. A line was fired from the tug, and men on the supertanker's decks caught it and tied it down as high seas repeatedly swept across the decks, threatening to heave them overboard. This line was connected at the tug's end to a much heavier cable that would take the strain of moving the supertanker. The process of pulling that cable across the water from the tug took a lot of time. By 1:30 P.M. it was firmly in place, but there was some delay because agreement had to be reached with the ship's owners in Chicago over the legal standing of the tug's captain. His rescue could be considered as either emergency help or a salvage operation, and large sums of money were involved in this decision. Two valuable hours were lost through these discussions before towing operations could continue.

As the tug began to move the tanker to point it westward so that it could start up its engines and move farther away from shore, the tow cable broke. Several failed attempts were made to attach a new cable to the ship. At 8:00 P.M. the captain decided he had to drop anchor. He felt that the ship was dangerously close to submerged rocks and hoped that the anchor would slow the shoreward drift. Then, as suddenly as the rudder had failed, a steam pipe broke, cutting off all power to the winches that would be needed either to pull a tow cable or raise the anchor. Close to midnight, at high tide, the *Amoco Cadiz* ran aground. Had it happened at low tide, it might have been possible to float the supertanker when the tide came in. Almost immediately, the ship began to lose oil as cracks developed in the hull. The loss of oil increased as wave after wave pressed the vessel against the rocks and the cracks in the hull became deep gashes.

CONSEQUENCES

On the morning of March 17, the reality of the spill was evident to all who lived along France's northwest coast. The smell of oil was everywhere, and people who ventured near the water experienced a stinging sensation in their eyes from the airborne fumes. Within the following few days, as the ship broke open, the rest of its 223,000 tons of crude oil was spilled. Much of the 200 miles of coastline from near Saint-Malo on the north to Nantes on the west was covered with oil. Subsequent calculations by French authorities estimated that 25 percent of the cargo was washed ashore while another 20 percent became embedded in the coastal sand. About 35 percent evaporated, and the rest lay in the water between the surface and seabed.

Oyster beds are a major source of income along this coast. There are 11,000 tons of them, and the immediate reaction of some owners was to collect as many oysters as possible and move them to a safer beach area farther south. For most of those involved in the oyster industry this was not possible, and their loss was catastrophic. Little was saved. It would take years for these oyster beds to be restocked. Tanks of lobsters that were being held in anticipation of the tourist season were also lost, and millions of small marine animals were killed. These included sediment-dwelling wildlife such as cockles and clams. Free-swimming fish that were able to escape the oil were less affected. The highest mortality occurred where an obstacle such as a bay trapped the oil, preventing it moving into open water. In one area of three square miles formerly rich in wildlife, 33 million dead marine animals were counted.

With persistent strong winds continuing for several days, the oil piled up on the beaches and rocks. It penetrated the sand to a depth of 20 inches on several beaches. Different layers were saturated because of the movement of sand as waves hit the shore. This proved to be a frustrating problem when cleanup began. Piers and slips in the small harbors were quickly covered with oil. Oil remained for only a short time along exposed rocky shores because they were under continuing pressure from powerful waves, but in sheltered areas it was a different story. Oil lingered there in the form of a crust of asphalt and stayed—in some cases for years.

Hundreds of thousands of tourists visit this coast annually, and the economic impact of losing them weighed heavily on the minds of the local business community. As the year progressed it became obvious that losses to tourism would be enormous. Sales of fish and crustaceans dropped $2 million in March 1978, $3 million in each of the following two months, and there were $9 billion lost in accommodations revenues before the year ended. This was not the end of the problem. The widespread publicity that accompanied the spill created its own phobia in the minds of people elsewhere. They thought that all of northwest France was covered in oil, and they did not want to eat anything from it. Even vegetables grown in the region could not be sold.

CLEANUP

A mile-long segment of boom (a floating narrow band of impermeable material) was brought in to protect the smaller bays, and with constant monitoring this worked reasonably well. The sheltered area was protected from high winds and excessive quantities of oil. Boom was largely ineffective in other areas due to strong currents and enormous quantities of oil. There, every attempt to prevent oil movement ended up with waves washing as much oil over the boom as it held back. Skimmers were equally inefficient. Pumps and hoses were blocked by seaweed. Vacuum trucks proved to be the best first approach for removing oil from piers where the seaweed was thick. These wagons were able to pump oil, water, and seaweed. After the water was separated from the oil, the tanks were emptied into interim storage tanks.

High-pressure hot-water hoses proved to be very effective in cleaning oil from the landward side of rocky shorelines. A small amount of dispersant was then applied to prevent fresh quantities of oil from reaching them in the next high tide. Oil collected from the shore via the vacuum trucks was stored at Brest, the biggest urban area in this region. Under the kind of weather being experienced at this time, oil forms into a viscous, chocolate mousse–like emulsion of water and oil that weighs much more than oil alone and is hard to remove. This greatly complicated the cleanup efforts. Of the oil that reached the shore, 149,000 tons changed into this emulsion. It weighed close to half a million tons, more than twice the weight of the spilled cargo.

The official plan for coping with oil spills was quite inadequate. It had been designed for spills of no more than 20,000 tons. Disillusionment with detergents in earlier spills meant that they could not be employed. The only effective course left was to clean the beaches. In France, that job is always assigned first to the army. All along the 200 miles of coastline they used the only tools that would work: shovels. Oil-soaked sand was bagged and carried away while village fire trucks washed soil from rock surfaces. Workers would clear a beach only to discover new coatings of oil as the next series of waves removed the top layers of sand. Oil would sometimes accumulate to a depth of four inches in harbors, then suddenly be blown out to sea when the wind changed.

Because of the large scale of the problem, the soldiers were assisted by detachments from the navy and the police. Altogether 6,000 military personnel were brought in for the work. Tankers collected the material taken from the beaches and transported it to a safe place inland. Often police escorts were needed to clear the roads of the sightseers who converged on the scene. Bulldozers scooped up good sand wherever it could be found and moved it above high water. Later, it would be placed back on shore. Because of the harmful effects of the oil fumes, workers had to wear goggles and gloves. Even so they suffered from nausea and headaches.

Eight weeks after the main cleanup had begun, the work was finished and most beach areas were free of oil. The longer a cleanup takes, the more oil seeps into the sand and grass. Many questions were asked in the wake of the disaster. Why did the navy, the agency responsible for responding to accidents in French waters, fail to help the *Amoco Cadiz*? Naval authorities replied that they received no reports of difficulties until it was too late. Why did the supertanker sail within 6 miles of land when international agencies such as the United Nations recommend a minimum distance of 15 miles from shore? The answer to that was that the inshore lane was the right one for incoming tankers because it minimized the risk of collisions with outward-bound ships.

The French government was unwilling to accept the necessity of huge tankers routinely passing so close to shore. It was determined to avoid another tragedy like the *Amoco Cadiz* spill. New traffic lanes were introduced, ensuring that ships would be more than 30 miles from shore instead of the existing 7. At the same time, new regulations were introduced and additional resources were provided for the navy, which was instructed to establish a 24-hour watch by sea and air. All tankers were required to report their movements while in France's territorial waters and to give details of any problems in their ships' operations. To assist the navy, two of France's largest oceangoing tugs were permanently based at Brest, the nearest point of contact with tanker shipping, and new radio beacons and a radar buoy were installed offshore.

As often happens in shipping, the *Amoco Cadiz* was registered in an African country in order to gain tax and insurance advantages for the owner. It was owned by Standard Oil of Indiana, and this company together with other oil companies had agreed on a figure of $30 million to cover costs arising from a spill. It was obvious to all that this figure was inadequate to cover the costs arising from the *Amoco Cadiz* accident. French companies lodged claims of $750 million in the circuit court of Cook County, Illinois, where Standard Oil was registered. Others parties, including the French government, added other claims.

In the legal actions that followed, a U.S. judge finally decided that Standard Oil was responsible for the accident and guilty on several counts, such as failure to train the crew adequately. The allocation of payments was quite a different matter. Some claims were duplicates of others, and some had been indefensibly inflated. In the final settlement, the companies that represented the 90 French communities that had suffered most were awarded $85 million, little more than one-tenth their $750 million claim. For the 400,000 people concerned, that amounted to less than $220 each.

Exxon Valdez
Prince William Sound, Alaska
March 24, 1989

At Prudhoe Bay, on the north coast of Alaska, large deposits of oil and gas began to be intensively exploited in the late 1960s. By 1977, an 800-mile pipeline was completed at a cost of $8 billion, bringing oil to ports on Alaska's south coast where the milder climate permitted year-round operations. From the moment the first big tanker arrived at Valdez, a principal oil-loading terminal, there were environmental concerns. The ship channels along this coast are narrow and rocky and weather conditions are

Exxon Valdez Oil Spill Trustee Council containing the spill in the water. (© National Oceanic and Atmospheric Administration/Department of Commerce)

Exxon Valdez Oil Spill Trustee Council cleaning up after the oil spill. (© National Oceanic and Atmospheric Administration/Department of Commerce)

often bad. Shipping channels are frequently clogged with blocks of ice that break off from glaciers. To make matters worse, tankers like the *Exxon Valdez* had single hulls, not the double hulls that would be safer in such dangerous waters.

Oil was shipped from Valdez to refineries in Washington State and other places along the west coast of the United States as well as to Hawaii. The volume transported was huge, with two or three supertankers arriving each day. Alaska was the second-largest U.S. producer of oil, after Texas, and it accounted for one-quarter of the nation's supplies. Responsibility for the safe transportation of oil rested with a consortium of several oil companies, called Alyeska, which built both the pipeline and the terminal at Valdez. Spills of varying amounts were frequent both along the pipeline and near Valdez, but it was the *Exxon Valdez* accident, because of the size of the spill, that heightened concern throughout the country.

CAUSES

The *Exxon Valdez* was a supertanker built in San Diego, California, in 1986 for the Exxon Shipping Company. It was designed for oil transportation from Alaska, and its safety features included separate compartments for oil plus all the necessary navigational aids required by United States Coast Guard regulations. On March 22, 1989, at 11:25 P.M. it arrived at Valdez for its next cargo. It had been loaded with ballast water to maintain stability on the trip north, and the emptying of this ballast, which began shortly after midnight, was completed at 4:15 A.M. on March 23. Loading the crude oil cargo began an hour later and was completed at 7:24 P.M. the same day. About an hour and a half later the *Exxon Valdez* was on its way out of Valdez, escorted by a pilot for the first part of the journey, through the Valdez Narrows.

The crew was very tired by this time. Shipping rules required that they be given several hours off duty before sailing, but the company ignored this rule in order to get the oil to market as quickly as possible. Another rule was also ignored: the crew must not consume any alcohol within four hours of sailing. Captain Joseph Hazelwood had been drinking within an hour of the ship's departure, which might explain his strange behavior shortly after 11:00 P.M. Both the captain and his chief mate, Kunkel, were on the bridge with the pilot as the ship sailed through the one-mile-wide Valdez Narrows. Then, after the pilot returned to Valdez, the *Exxon Valdez* headed out into Prince William Sound. Large quantities of ice had drifted across the main channel from the Columbia Glacier, and the tanker took an eastward course away from the main traffic lane to avoid the ice after receiving permission to do so from the operator controlling all movements of ships in the area.

The video tracking control system (VTR) in place to monitor and to control the movements of ships was not required to maintain watch on the *Exxon Valdez* after it left the Valdez Narrows. A memorandum issued by the port authority two years earlier had eliminated that requirement. Beyond the narrows it was left to the discretion of operators whether or not to check on the locations of ships. In the case of the *Exxon Valdez* no attempts were made to do so. Had the systems followed it, the ship's fatal movement toward Bligh Reef would have been detected and a safe alternative taken.

As the vessel entered Prince William Sound, Captain Hazelwood made two dangerous decisions. He placed the ship on automatic pilot, a move that would prevent a helmsman's ability to make quick moves in an emergency, and then told his third mate, Gregory Cousins, to take over control of the ship while he retired to his cabin. Cousins had neither the training nor the experience to take a ship through ice-covered waters; it was the captain's responsibility to be on the bridge when navigating this narrow channel between the ice and the reefs nearby. As the vessel approached Bligh Reef a flashing red light on a buoy beside the reef became visible on the right side of the ship. It would have been on the opposite side if the *Exxon Valdez* had been on the right course.

Cousins immediately ordered the helmsman to swing the ship 10 degrees to the right, but nothing happened. The ship continued straight ahead. Cousins tried again with the same result. Only then did he realize that the captain had put the ship on automatic pilot without telling him. He rushed to turn off the autopilot and shouted to the helmsman to swing right 20 degrees. The vessel once again failed to move—this time for a different reason. It had hit the sea floor and a few minutes later was stuck fast on Bligh Reef. Oil gushed from the cracked hull.

The final report of the National Transportation Safety Board (NTSB) listed a number of causal factors. High on its list were the tiredness of the crew and the captain's inebriation. Tiredness or the lesser abilities of those on the bridge led to the initial failure to see a red warning light near the reef. It flashed for several minutes. Likewise, exhaustion probably led to the third mate's failure to act quickly enough after discovering that the ship was on autopilot. The Exxon Shipping Company was blamed for not adequately monitoring Hazelwood's drinking problem and for failure to take account of the increased workload of a crew that was smaller to save costs.

The NTSB concluded that the main causes of the accident were the captain's failure to provide a proper navigation watch because of his alcohol impairment and Exxon's failure to provide a fit captain and a crew that was not overworked. In addition, it pointed to the weakness of the traffic-monitoring system, inadequate personnel training, and deficient management oversight as contributing factors.

CONSEQUENCES

Hazelwood, who seemed dazed and unconcerned, thought he could pull the ship away from the reef, but after six attempts he gave up. When a Coast Guard officer who arrived two hours later tested the captain, he was found to have a blood alcohol reading well above the permissible level.

Oil continued to pour out of the *Exxon Valdez*, accelerated by Hazelwood's attempts to move the ship. Hazelwood's decision to leave the bridge shortly after receiving permission to alter course was later considered a serious mistake. It was also contrary to federal regulations and Exxon policy. The maneuver around the ice required an intricate series of movements that demanded careful timing and judgment. Frequent fixing of the ship's position was essential because the space between the glacial ice and the reef was very small for a supertanker. These decisions should not have been made by less-qualified people.

Damage came to $25 million for the vessel and $3.5 million in lost oil. For days the oil leaks continued. Finally, with the aid of another ship, 80 percent of the cargo was taken out of the *Exxon Valdez*, leaving 11 million gallons scattered along shorelines for 150 miles and spread over thousands of square miles of ocean surface. In the weeks that followed, an additional 300 miles of beaches to the west of Kodiak Island were smothered in black oil. The problem was more serious than it would have been in more southern latitudes because in the colder northern waters of Alaska the contamination lingers. The ocean takes a longer time to absorb the oil.

First concerns centered on the marine- and shore-life populations. The number of species affected was enormous. One report listed 120 forms of life, some of them vital to Alaska's economy. They included salmon, otter, herring, halibut, whale, shrimp, sea lion, Arctic fox, Arctic loon, and bald eagle. Everywhere the oil came ashore the dead and the dying were visible among the myriad life-forms, from tiny plankton to big sea lions. The delay in the cleanup worsened the situation. Crude oil, which is lighter than water, has many toxic elements, and with the passage of time certain changes occur. The oil can adhere to suspended particles and sink, thus damaging biological life on the seabed. Mousselike emulsion can form, creating difficulties for skimmers and preventing burning the oil.

The time of the spill was early spring. Salmon fry were emerging from gravel beds in freshwater streams, herring were returning to spawn, and sea otters, seals, and sea lions were giving birth to their young. Native communities were the hardest hit. Their subsistence lifestyle depended on seals, herring, clams, and crab, and most of these were destroyed by the spill. Nothing in their long history prepared them for the sudden disappearance of their traditional food resources. Commercial salmon and herring fisheries were equally devastated. One long-term effect was vividly demonstrated five years later when 100 fishing boats blockaded the entrance to Valdez Narrows for two days because the seasonal salmon run had plummeted. The fish that might have appeared were the ones that tried to reach the ocean at the time of the spill.

There was a different kind of concern regarding migrating birds and mammals. Prince William Sound is the world's single largest stopping place for waterfowl. In summer this area is also visited by large numbers of humpback and killer whales. Estimates of the numbers of wildlife that died include 5,000 sea otters, 300 harbor seals, 20 killer whales (almost all females with calves), and 14,000 marbled murrelets among the half-million dead birds.

CLEANUP

Damage to human health took its own toll. Cleanup workers faced average oil mist exposure 12 times in excess of regulatory limits, with a maximum exposure 400 times higher during hot-water beach washing. Alyeska's first reaction was to send a barge carrying dispersants to help break up the oil particles and thus make it easy for water to absorb them. That happened 14 hours after the start of the spill and tests were immediately conducted on their use, but before long they were cancelled as failures because of bad weather. Europeans had earlier tried the same approach and had given it up for the same reason. In Alyeska's procedural manual for dealing with a spill of the size of the *Exxon Valdez* one, the overriding consideration was that a spill of that magnitude is very unlikely to occur. Recommended steps were to send a tug to the scene of the spill with dispersant and other equipment within five hours, but even that scenario did not materialize. Even if Alyeska had responded, there were not enough skimmers or boom or dispersant available to do an effective job.

A trial burn was attempted using a fire-resistant boom that was tied to two towlines and moved slowly through the main portion of the slick until it filled with oil. The boom was then towed away from the slick and the oil was ignited but, as with dispersants, bad weather made the process inoperable. Mechanical cleanup was started using booms and skimmers, but thick oil and heavy kelp tended to clog the equipment. Repairs were time-consuming. Transferring oil from temporary storage vessels into more permanent containers was also difficult because of the oil's weight and thickness. It was increasingly clear that the influence of bad weather had not been given adequate consideration in Alyeska's cleanup plans.

After the various delays as various possible approaches were tried, a more coordinated cleanup program was laid out. Exxon was one of the biggest contributors to the plan. It hired thousands of workers to start cleaning up the beaches. One of the greatest concerns of local fishers was the danger of hatcheries' being damaged, as this could destroy their livelihood for years. They formed a single work group and succeeded in keeping the spreading oil slicks away from their hatching beds. Over the course of the summer the beach-cleaning workforce increased to 15,000 but, even with all their efforts, the beaches were still dirty as fall arrived.

One year after the spill only a quarter the normal volume of birds and whales appeared. Ten years later, only 2 of 26 species examined in detail—bald eagles and river otters—seemed to have made a complete recovery. Only one-seventh of the oil was removed during cleanup operations. The rest became embedded in beaches, national parks, wilderness areas, and the ocean floor.

Despite the tragedy of the *Amoco Cadiz* 11 years earlier in Europe, little had been done to strengthen Alaska's readiness to deal with a spill. When the U.S. Congress initially approved the building of a pipeline and the Alyeska terminal in 1971, strict regulations were put in place for dealing with possible spills: Alyeska was required to have appropriate resources on hand to tackle oil spills immediately, and oil companies were to be held responsible for accidents caused by their tankers. When the *Exxon Valdez* struck the reef, neither the terminal nor the shipping company was able to act. The former had no booms, skimmers, or other supplies on hand, and the latter had never been plainly told that it was responsible.

The initial pipeline and terminal approval by Congress in 1971 recommended the use of double-hulled tankers, and Alyeska promised to do so. No double-hulled tankers appeared after 1989, and even after Congress passed a law in 1990 requiring them, none were put into commission within the first 10 years. Oil companies managed to persuade both federal and state authorities that they were not yet practicable.

Alaska passed a bill in 1990 requiring the oil industry to stockpile enough supplies to deal with a 12 million–gallon spill. Other bills dealt with reimbursement of people victimized by spills; monitoring of tankers, pipelines, and other vessels at sea; and penalties for violation of the rules governing tankers. The U.S. Congress took action by raising the liability tonnage level for shippers to eight times its previous amount. Congress added a new five-cents-a-barrel fee to provide a $1 billion oil-spill response and cleanup fund. Ten years after the spill and five years after a jury ordered Exxon to pay $5 billion in punitive damages, the debate over costs and who should pay was still being debated in the courts.

North Cape
Block Island Sound, Rhode Island
January 19, 1996

The stretch of water along Rhode Island's southern coast is the regular path for all coastal traffic to and from New York. At any given time huge volumes of oil pass near the shore, and people around Rhode Island Sound are very much aware of this because the coastal area is a natural-resource community; that is, life there is centered on the biological and natural

The tug *Scandia,* grounded in a winter storm, was pulling the *North Cape.* The barge carried approximately 828,000 gallons of home heating oil when the tug and barge grounded in Rhode Island waters. (© National Oceanic and Atmospheric Administration/Department of Commerce)

environment. Both economic and recreational activities depend on renewable natural resources. The fishing industry alone is worth $500 million annually. In summer, people come from all parts of the state to enjoy the area's recreational facilities.

CAUSES

The tug *Scandia* and an unmanned barge carrying 3.9 million gallons of home heating oil, the *North Cape*, were on a coastal trip from New York to Providence, Rhode Island, on January 19, 1996. Intense southerly winds were forecast that day. While traversing the relatively protected waters of Long Island Sound, the tug apparently experienced some engine problems. Continuing eastward, the *Scandia* and the *North Cape* reached the unsheltered stretch of Block Island Sound where by midafternoon the southerly winds had reached 60 miles per hour and waves were as high as 20 feet. Both tug and barge were pushed toward the shore, with the tug unable to counter this movement.

A fire broke out in the tug's engine room when it was a few miles from Point Judith. After sounding a distress call, the crew abandoned the tug, which was soon engulfed in flames. The United States Coast Guard responded and rescued the crew from the water. At that point, both barge and tug were drifting toward the coast under pressure from wind and tide. Two men went on board the barge during the late afternoon and attempted to anchor it. That effort failed because the anchor could not be released, and the men were taken off. Wind and wave conditions worsened. By early that evening the tug and the barge were aground on a beach adjacent to the Ninigret National Wildlife Refuge, a place that attracts 75,000 birds, including rare harlequin ducks. Damaged from the grounding, the barge spilled 828,000 gallons of heating oil that rapidly spread to the coastal ocean. Fundamental causes of the accident were twofold: poor maintenance of tugs, and lack of care when transporting dangerous cargoes in bad weather.

CONSEQUENCES

Within hours of the spill, dead marine organisms and sea birds began washing up on the beach. The next morning oil was visible on the water. In the following days, the magnitude of the disaster was evident. Oil spread from Block Island Sound toward Rhode Island Sound and Narragansett Bay, forcing a closure of fishing for an area of 250 square miles of ocean. This was followed by other closures. Ponds with lobsters, starfish, and clams were all shut down, and a general slump set in for all segments of the fishing industry. In less than a month recreational fishing declined by 80 percent. The barge and tug were finally removed, but the damage had been done. Heating oil is lighter than crude oil and its long-lasting effects are accordingly less. In the short term, however, it is much more toxic and so damage to Rhode Island was extensive and deep.

Nine million lobsters had been killed by the spill, as well as 19 million surf clams. More than 4 million fish died, and more than a million pounds of marine life was lost—worms, crabs, mussels, and starfish. In the salt ponds, millions of worms and other amphipods had been killed and a total of a million crabs, shrimp, clams, and oysters lost. Birds were also harmed by the spill, with 2,300 killed, including 402 loons. The habitat of the piping plover was destroyed, and its restoration was given high priority in the final settlement.

CLEANUP

New regulations and a new awareness of the seriousness of any oil spill made it much easier in the aftermath of this spill to rehabilitate the damaged environment and to do it in less time. Within two years—a much shorter time span for final restoration than was possible with either the *Amoco Cadiz* or the *Exxon Valdez*—the National Oceanic and Atmospheric Administration (NOAA) had a detailed plan. It was financed by $16 million from the barge owner and others who had been responsible for the tragedy. Details of the restoration plan were based on research studies that were launched almost immediately after the spill.

Research findings had established that while a great deal of short-term injury was caused to natural resources, there were few or no permanent impacts. They also found that measures can be undertaken to restore the short-term loss of natural resources. This whole operation was the first use of NOAA's new damage-assessment regulations in relation to the Oil Pollution Act of 1990. Those responsible for Rhode Island's natural resources used the new regulations to determine how living and nonliving natural resources, including birds, fish, other marine biota, beaches, sediments, and water, were impacted by the spill and how they could be restored in the most effective way.

As part of the same overall agreement with the state of Rhode Island, more than a million female lobsters were to be added to Block Island Sound by 2005. These would gradually be purchased and introduced over time, and all of them would have their tails notched with a V. It is illegal to harvest lobsters so marked. As a result, the number of adult females and their offspring will increase rapidly over the five-year development period. To improve fish runs, obstructions on rivers or streams were removed to allow easy access to salt ponds. For the recovery of the shellfish industry, 10 million clams were added to specific habitats in Narragansett Bay and some coastal salt ponds. In the restoration of loons' habitats, arrangements were made to monitor nesting sites. In addition, a conservation site was purchased in nearby Maine where loons are known to nest.

SELECTED READINGS

Fairhall, D. and Jordan, P. *Black Tide Rising: The Wreck of the Amoco Cadiz*. London: Andre Deutsch, 1980.

Fingas, Merv. *The Basics of Oil Spill*. 2nd ed. Edited by Jennifer Charles. Boca Raton, Fl.: Lewis Publishers, 2001.

Keeble, John. *Out of the Channel: The Exxon Valdez Oil Spill in Prince William Sound*. New York: HarperCollins, 1991.

McLoughlin, William. *Rhode Island: A Bicentennial History*. New York: W.W. Norton, 1978.

7

Terrorist Acts

Anthrax Attacks
Eastern United States
Fall 2001

Anthrax is a disease that mainly affects grazing animals such as sheep, cattle, and goats. It was known 2,000 years ago, and in the nineteenth century vaccines were developed to prevent infection with the disease.

Throughout the twentieth century various governments conducted research on anthrax because of its potential use as a weapon, but it was never used in warfare. An accidental release from a research facility in the former Soviet Union killed 70 people, and in 1993 a secret society in Japan attempted to release anthrax spores in downtown Tokyo. That was the first time that the dangerous bacillus was directed against a civilian population.

The anthrax bacillus is a tiny organism. Hundreds of them laid end to end would measure less than one millimeter. When exposed to air, anthrax forms spores, and these can reproduce in the body of the host that receives them. Anthrax is not a contagious disease. Skin anthrax can usually be successfully treated with antibiotics, but the danger is far greater if the spores are inhaled: very few people survive unless antibiotic treatment is given within a day or two. In the vast majority of cases, the victims are dead within a few weeks. Anthrax is found in soil all over the world in most countries, and instances of human infection are frequently reported

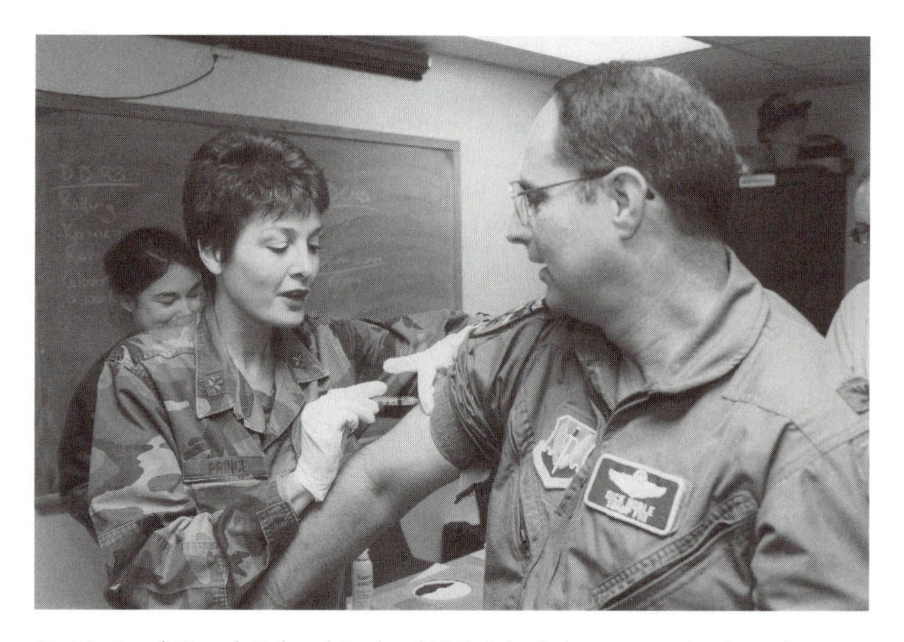

Air National Guard Colonel Richard Nobel (right), receives the first of a series of anthrax inoculations from Major Theresa Prince (left) at McEntire Air National Guard Station, SC. Noble volunteered to be the first member of the 169th Fighter Wing to receive his anthrax inoculation. Noble is the vice commander of the 169th which files the F-16C Fighting Falcon. Prince is chief of nursing services of the 169th Medical Squadron at McEntire. (© DoD photo by Senior Airman Marvin Preston, U.S. Air Force)

in Middle Eastern countries and in Africa. During the 1991 Gulf War, U.S. soldiers were inoculated with a vaccine to prevent their catching the disease because there was some fear that Iraq might use anthrax as a weapon.

CAUSES

In mid-October 2001, a few weeks after the devastation of September 11, when the World Trade Center towers were destroyed and the Pentagon was hit, letters containing anthrax spores began to arrive at various U.S. media centers and at government offices in Washington, D.C.

CONSEQUENCES

A photo editor at a Florida news agency was the first to be affected. He opened an envelope that arrived on October 15 and unknowingly inhaled some of the anthrax spores that fell out. Several days passed before the contents of the envelope were tested and identified. By then it was too late to do anything for the photo editor. He died two weeks later. Five days after the delivery in Florida the next anthrax envelope arrived, this time at the Washington, D.C., office of U.S. Senate leader Tom Daschle. Over the following two weeks several more anthrax letters arrived at media centers and various offices in Washington. There were also hoax letters that looked like the lethal ones. Most of the attacks occurred between October 15 and October 31. Places in at least nine states became involved.

As long as each mailing affected only the person to whom the letter was addressed, there was limited concern because care in handling mail and immediate treatment with antibiotics could cope with the dangers. Before long, however, it became clear that the trail from sender to recipient could itself be a source of contamination. On October 31, a postal machine sent to Indianapolis, Indiana, was found to have traces of anthrax. At the same time, anthrax spores were discovered in empty mailbags in Missouri. A general fear developed that the mail service might be infecting people and places everywhere it operated, both at home and abroad. Confirmation came in the form of new infections in people who did not work in any of the affected places.

By the end of October 2001, five people had died from anthrax, including two postal workers and a journalist, and thirteen others were ill. The postal system as well as several government offices were seriously disrupted. Thirty congressional workers had tested positive for anthrax, and the Hart Senate Office Building, the offices of the House of Representatives of the U.S. Congress, and even the Library of Congress all had to be evacuated at times because of the presence of anthrax spores. The mail facility in Fairfax, Virginia, which handles all the mail for the District of Columbia, was found to be contaminated with anthrax spores.

CLEANUP

Interim measures were taken in all government offices and in various places across the country. Many school and public libraries were closed. Several affected postal stations were also closed. The White House mail was quarantined and several government offices locked in order to check for spores while their staffs met elsewhere. For the first time in its history, the Supreme Court convened away from its own chambers. The State Department cut off all mail to its 240 embassies and consulates worldwide. One expert in government administration said that nothing since the days of the American Civil War had as seriously disrupted the business of government as this anthrax attack. He added that the economic losses would probably amount to billions of dollars.

As intense investigations into the source of the scourge continued, the U.S. Postal Service began the process of irradiating all mail at numerous locations. At the same time, extensive tests were conducted on the anthrax powder to find out how it was made and what type of anthrax was used. It was known that identical material had been employed in all the attacks, proving that the campaign originated from a single source and possibly from a single terrorist. Over the following six months, it was established that the powder used was quite different from that used by government biological defense programs in both the United States and other countries. This seemed to confirm Federal Bureau of Investigation (FBI) suspicions from the beginning that the whole campaign was a U.S.-based plan, not the work of terrorists from overseas. Research workers also discovered that the anthrax had been coated with a chemical substance that would prevent the tiny spores from clumping together so that they would float freely into the air once a letter was opened and inhalation would be likely.

Evidence of this type suggested that the person behind the attack was well acquainted with the chemistry and technology of using anthrax as a weapon. For instance, he or she might have worked in the U.S. government's biological defense program at some time and become disgruntled and was seeking revenge or notoriety. The author of the attacks certainly gained the reputation of causing the worst acts of biological terrorism ever known in the United States. FBI investigations are continuing, seeking data on genetic signatures that might help them localize the specific site where the anthrax originated. The FBI's earlier speculation that the culprit was a U.S. scientist with special training and skills has been strengthened by all the findings to date.

Oil Inferno
Kuwait and the Persian Gulf
1991

At the beginning of August 1990, Iraqi troops invaded Kuwait in order to gain control of its oilfields and make it a province of Iraq. The United Nations immediately condemned this action, and when diplomatic efforts to solve the crisis failed, a coalition of many countries was assembled to reclaim Kuwait by force. Air attacks on Iraq were launched early in January 1991, and ground forces later crossed into Kuwait. The ground war was

A destroyed Iraqi tank rests near a series of oil well fires during the Gulf War, Saturday, March 9, 1991, in northern Kuwait. Hundreds of fires continued to burn out of control, casting a pall of toxic smoke over the surrounding area and raised health and environmental concerns. (AP/Wide World Photos)

brief, and by the end of February 1991 the Iraqi military had been completely defeated. There were many Iraqi deaths, perhaps as many as 100,000. Among the coalition armies, between 200 and 300 were killed. This might have been the end of the story, but Iraq decided to launch a series of environmental acts of terrorism as it withdrew from Kuwait.

CAUSES

A flood of oil was released into the Persian Gulf, destroying most forms of life there. At the same time, hundreds of oil wells were set on fire within Kuwait, creating a massive blanket of air pollution (see figure 7.1) From the air, the country looked like a huge black blanket through which oil flares shot upward from time to time. It was a catastrophe with great implications for the future of the surrounding environment, comparable in its destructiveness to the Chernobyl nuclear disaster and greater in its extent than any other oil spill in history.

CONSEQUENCES

Animals and people alike had trouble breathing. There was an unpleasant smell everywhere that permeated clothing and stung as it irritated lungs and skin. On the ground lay pools of oil that caught fire occasionally

Figure 7.1
Area Affected by the Oil Well Fires (Paul Giesbrecht)

as some nearby flame reached them. Trees, buildings, cars—anything on the land's surface—were covered with tar.

During their short period of occupation, the Iraqis had stripped the country of everything movable. Roads had to be created from fire site to fire site because the soldiers had cut defensive trenches across highways. In addition, the retreating army had left stores of ammunition and discarded vehicles everywhere. Minefields had to be cleared, but no one knew where they were.

The scale of destruction by the Iraqis was so great that every innovative method possible to extinguish the fires was welcomed. Each day, the equivalent of about 15 percent of the world's daily consumption of oil was going up in smoke or forming rivers of oil. That amounted to six million barrels, roughly the quantity consumed daily by all the gasoline-powered vehicles in the United States. This loss went on for almost two months before the first fires were extinguished. Two months' supply is often the amount of emergency reserves stored in Western countries. Furthermore, Kuwait was not the only casualty of Iraqi's environmental terrorism. All the surrounding countries, covering an area twice that of Alaska, were smothered with poisonous black air, creating all kinds of health problems.

CLEANUP

Specialists in fire control were brought in at an early stage from all over the world. So challenging did the task seem to them that their first estimate for putting out all the fires was five years. Logistical problems faced them on every hand. The airport was not accessible, so they had to wait for the smoke and fires around it to be cleared before they could bring in personnel and materials. The fire crews used huge bulldozers to pile up heaps of sand to fill in the trenches. At the same time, they were able to use these mountains of sand to absorb the impact of exploding mines and thus get rid of them.

Most of Kuwait's oil wells are operated by underground pressure. There is no need for the kind of surface derricks so common in North America, where oil has to be pumped up. Each Kuwaiti well is therefore a small, inconspicuous structure carrying the usual Christmas tree–like structure of control pipes and surrounded by a chain-link fence. Where Iraqi damage was minimal, it was possible to stop fires by using an ordinary wrench. These opportunities were few, however. For the most part, explosives had destroyed the control equipment. A critical part of the controls was the blow-out preventer, a valve that adjusts pressure to cope with sudden increases from below. The loss of this control as fires were being extinguished led to sudden ignition in a few places, killing workers. Five men lost their lives in this way.

For the bigger fires, a remote-operated crane was used to place a huge, wide-diameter pipe over the well. Water and dry chemicals were then poured into the pipe to smother the flames. Both the heat and the noise made it impossible for workers to talk to one another when they were close to the wells, so hand signals were used. Large supplies of liquid nitrogen were on hand in Saudi Arabia, and this was found to be an excellent chemical for putting out a fire. Its minus 320 degrees Fahrenheit temperature was an ideal cooling agent. Where flames shot out of a well horizontally as well as vertically, it was too dangerous to use the wide-diameter pipes. Explosives were the only answer. They quickly put out the flames, and although they caused additional damage, they made it possible to reach the well and rebuild it.

As each team coped with one fire after another, the effect on the project as a whole was quite dramatic. Visibility gradually opened up. Teams could see better what was going on and were able to tackle those fires that seemed to be doing most damage. Instead of the original five-year prospect, the end of their work began to look more like one year. The final landscape was a wasteland. Swamps were everywhere, mainly filled with a mixture of oil and mud, and the earth had become an oil-saturated, black mass. It would be a long time before any plants took root in that kind of soil. The end came in November 1991 when the last oil-well fire was put out eight months after the first foray.

The other half of the environmental catastrophe, the oil flooding into the Persian Gulf, was receiving the same intensive attention as the flaming oil wells over the same period of time. In some places nothing could be done. Salt marshes, mangroves, and coral habitats of rare sea turtles were destroyed. Estimates of seabird deaths reached 30,000. Fishing is a major industry all along the shores of the gulf. After oil, it is the main source of income for thousands of people. All their fish stocks, including shrimp, barracuda, and mackerel, were wiped out.

Forty times the amount of oil spilled by the *Exxon Valdez* in Alaska was released into the gulf, adding to an earlier spill. In the 1980s, when Iraq and Iran were at war, Iraqi missiles hit offshore Iranian oil platforms and spilled two million barrels of oil into the ocean. A decade later, the new flood of oil was attacked with booms and skimmers, and one million barrels of oil were recovered from the gulf's surface. That was a record for any spill, and it was encouraged by the unique demands of Saudi Arabia's desalination plants.

Three of these plants are on the gulf side of Saudi Arabia, and they provide 40 percent of the nation's drinking water. It was the top priority of the Saudi government to prevent any oil from reaching the intake pipes of these installations. One plant alone produces 220 million gallons of fresh water a day, supplying 75 percent of the water needed by the capital city, Riyadh. That same installation is also the source of water for a range of industrial enterprises in and around Riyadh. All the intake sites to these desalination plants were surrounded with several lengths of boom as soon as news of the oil flood reached Riyadh. The booms were arranged in an inverted V formation to deflect oil away from the plants and to minimize the risk of oil splashing over the booms.

The final cost of all the cleanups was more than $12 billion (including the value of lost oil), but there was little prospect of collecting these costs from the people responsible. Any national leader who could do what Iraqi president Saddam Hussein did in and around the gulf region is not likely to recognize any claims. Several international conventions exist for dealing with the Kuwaiti catastrophe. They extend from the Hague Convention of 1907 condemning warring nations for environmental destruction to similar agreements in 1949 and 1977. Economic sanctions against Iraq were the only possible response by the international community in relation to these conventions and agreements, and they were imposed at once. Ten years later they were still in place until Iraq was invaded and occupied in 2003.

Pentagon Washington, D.C. September 11, 2001

On Tuesday morning, September 11, 2001, there was a major terrorist attack on the United States. Two commercial jetliners crashed into the World Trade Center's twin towers in New York City and a third hit the Pentagon in Washington, D.C. A fourth, which many people believe was headed for the White House, crashed in Pennsylvania when passengers, at the cost of their lives, fought the hijackers but were unable to take control of the plane.

The Pentagon, as its name suggests, is a five-sided building constructed in five concentric rings of offices around an open center. Each ring is five stories high. On any given day 24,000 people work there. It was built in the 1940s during World War II and was being renovated to provide protection against terrorist attacks. In the renovation process, shatter-reducing Mylar had already been installed in one wing to absorb some of the shocks from an explosion and blast-resistant windows were in place. By September 11, this was the only part of the Pentagon that had been renovated, and workers had not yet moved into it. By good fortune it happened to be the part of the building that was hit by the terrorists, so casualties were few.

CAUSES

The airplane that hit the Pentagon began as American Airlines Flight 77, which left Dulles International Airport outside Washington, D.C., bound for Los Angeles, California. It took off at 8:00 A.M. with 64 passengers and crew aboard, including five hijackers. Soon after departure it stopped responding to air-traffic-control messages and was observed to turn around and head back to Washington, D.C. Shortly before 10:00

A.M. it slammed into the lower floors of the Pentagon, clipping trees and lampposts as it descended and plowing through floors one and two as far as the third concentric circle.

CONSEQUENCES

The Pentagon shook violently as it was hit. Workers inside heard the roar of the impact and almost immediately saw the outburst of fire and smoke. Lights went out as electricity was cut off. Even if lights had remained on it would have been almost impossible to see for any distance. Black smoke smothered everything. Flames threatened people as they made their way along corridors seeking an escape route. Pools of aviation fuel ignited from time to time, adding to the general confusion. Vehicles posted nearby also caught fire. Emergency procedures immediately came into operation, and every effort was made to get people out of the building. About 15 minutes after the plane crashed into it, the entire renovated segment collapsed. Arlington County, Virginia, police and fire departments were on the scene in less than an hour, tackling the fires and establishing order. Helicopters and ambulances took the injured to hospitals.

Special care was taken to safeguard classified documents. The National Military Command Center, which is located in the Pentagon, had been affected. Despite problems with smoke, it continued its work, with officers monitoring military operations worldwide. An early casualty was an overloaded cellular phone network, and those who were fortunate enough to get through took names and phone numbers from others and relayed them to their contacts with the request that they follow up with calls from their own phones. By 8:00 P.M. on September 11 fires were still burning strongly and spreading to new sections of the building. Core materials in the layered roof that had caught fire were particularly difficult to extinguish. Total costs of the whole tragedy were $740 million. Altogether 126 lives were lost on the ground, plus 64 in the plane.

CLEANUP

Six months after the attack all traces of the original damage were gone and much of the reconstruction had been completed, with one-third of the workers displaced by the events of September 11 back at work. The damaged segment of the Pentagon was repaired, including the renovations that had been in progress at the time of the attack, and a time capsule was placed in the outer wall of the building at the point where the terrorists had struck. The capsule contained the names of all who were killed in the attack and a collection of objects representing the time and place of the attack. No date was suggested for its opening in the future. The planned improvements were also accelerated in other parts of the building.

World Trade Center Attack
New York City
February 26, 1993

The World Trade Center (WTC) included the landmark twin towers of 110 floors each located on a 16-acre site near the southern tip of Manhattan Island. They rose more than 1,350 feet above street level, and in 1970, when they were first occupied, they were the world's tallest buildings. Their elevator system was a combination of express and local elevators, an arrangement that increased the amount of floor space given to occupancy. In conventional systems, only 50 percent of the area on each floor is available for offices. In the WTC it was 75 percent. Economy of space was obtained by having three vertical zones, from the ground floor to the 41st floor, then to 74th, and from there to the top. Express elevators served the three zones, and four banks of local elevators operated within each vertical zone.

The crater left in an underground parking garage following the 1993 World Trade Center bombing. (AP/Wide World Photos)

CAUSES

Just before noon on February 26, 1993, an explosion occurred in an underground garage beneath the WTC, powerful enough to rock the towers and demolish the steel-and-concrete ceiling of the underground train station, a major transportation point for New Jersey commuters. The bomb was located where it could do maximum damage.

CONSEQUENCES

A huge hole was ripped in the train station's wall and an even bigger cavity was created beneath. Thick black smoke from a fire that erupted swept upward to the top of both buildings, in which as many as 100,000 people worked or visited daily. On February 26 there were 50,000 people in the buildings, including 200 visiting kindergarten and elementary schoolchildren. Power for the entire complex was cut off, leaving people without lights, heat, or elevators.

Visitors to the towers had to be left for hours on the observation deck until injured people were attended to. To the thousands in the building it was a terrifying experience. Hundreds of people poured out of the towers into the streets, their faces black with soot, some of them having managed to find their way down from as high as the 100th floor. Many others stayed on their floors waiting for assistance to arrive. They packed cloths against doors and vents where smoke was entering or used moistened cloths on their faces. The blast was the equivalent of several major multialarm fires combined. Many fire department units from other parts of the city had to be called in to help. The statistics tell the story well. Six people died and more than a thousand were injured, 15 of the latter having received traumatic damage directly from the blast. Eighty-eight firefighters and 35 police officers were injured.

One newscaster unfortunately went on the air and advised people in the towers that if they were having trouble breathing, they should break out the glass windows. This was the worst thing he could have said, and his call was quickly refuted by others. There were more than 500 emergency personnel on the ground who could be hit with flying glass and, furthermore, open windows would allow smoke to enter the area. Charles Maikish was the director of the WTC, and as he felt the tower sway a little beyond the normal he knew that a major accident had taken place. He was on the 35th floor at the time, so his first move was to check the elevators. They had already moved back down to their starter floors in accordance with emergency procedures.

CLEANUP

Maikish made his way to the lobby and began to organize a command center there. Smoke was everywhere. A police officer on duty in the building broke a hole with his bare hands in the roof of an elevator and consoled a group of five-year-olds who were stranded in it. A fire fighter broke down an elevator door and found it full of semiconscious people lying on the floor. Down below in the parking area, people arriving at the time of the blast witnessed the smoke and fire and heard the screams of those who were closest to the bomb.

New York's television stations were located on the top of the first tower, and broadcasts by all but one were cut off. The operating station did not depend on the towers for power, so it and various radio outlets provided information for people inside the towers. Everyone was urged to stay calm and to wait. The scale of the rescue effort was huge, with rescuers having to reach and help people on 110 floors in each of the two towers and not knowing how many were stuck in elevators. It was impossible to say how long it would take to get everyone to safety, and meanwhile anxious relatives and friends waited below. For two hours the fire and smoke persisted. A number of disabled people had to be rescued by helicopter from the roof. That night the last elevator was reached close to midnight. Several people had been stuck in it for 11 hours. The terrorists had planned to send up a cloud of cyanide gas amid the smoke of the fire, but the cyanide was burned up in the heat of the explosion and did not vaporize.

The chief of the New York City Fire Department provided a summary of the events of February 26. He pointed out that the department had made numerous trips to the WTC since 1970, when it was first occupied. These related to minor fires, fire alarms, and one or two major fires, but nothing in these experiences prepared the department for the events of February 26. It was the largest incident ever handled in the city's 128-year history.

Approximately 25,000 people had been evacuated from each tower. Most of the victims were trapped on the upper floors, hence the large amount of time needed to rescue them. The bomb weighed more than 1,000 pounds and did comprehensive damage to seven floors, six of them below street level. The crater it made measured 130 by 150 feet and was located beneath the Vista Hotel. While the emergency work was concentrated in one day, the Fire Department maintained a presence at the WTC for a further month.

Arrests of four of the six terrorists—those who were still in the United States—came quickly because the Federal Bureau of Investigation had an informant who taped conversations with them two months after the bombing. Their trials were held in New York, and they hired the best lawyers they could find. William Kunstler, a well-known defense attorney, represented four of them. The trial took six months, during which the jury had to be kept together with protective security. All of the rights of the accused were safeguarded as fully as they would be for any American.

The judge handling the court cases knew that he was dealing with religious extremists for whom neither justice nor life had much value. They had their own view of Islam, and believed that to die in the name of Allah was a holy act. The jury found all four of them, Mohammed Salameh, Nidal Ayyad, Mahmoud Abouhalima, and Ahmad Ajaj, guilty. Pandemonium broke out as soon as the verdicts were given as people shouted out God's name and vented their anger at what they called injustice. Salameh, who thought he had won the case against him, lunged at the members of the jury and had to be restrained by marshals. As they screamed and hurled abusive language, the terrorist were handcuffed and dragged away, sentenced to serve 240 years each.

Before the leader of the terrorist cell, Ramzi Yousef, was finally caught and imprisoned, he had worked out plans for additional attacks. He came to the United States prior to the events of February 26 on an Iraqi passport and left very soon after the bombing. Then, in Manila, the Philippines, a fire broke out while Yousef was mixing some bomb-making material in January 1995 and he was forced to flee to avoid detection. He knew that U.S. authorities were on his trail. When investigators examined the site in Manila, they found evidence that led to his arrest in Pakistan a month later.

Yousef's plots were the most ambitious terrorist conspiracies ever attempted against the United States until the devastating events of September 11, 2001. The last of Yousef's five co-conspirators, Eyad Ismail, who drove the lethal truck to the WTC and then escaped after lighting the fuse, fled to Jordan, where he was captured in 1995. No one among the six is ever likely to be released. Yousef and Ismail also received prison sentences of 240 years. In the event that any of them tries to make money by publishing a book on the bombing, the judge levied fines of at least $10 million each to pay for restitution. Yousef was fined $250 million in damages for restitution. The actual cost of the damage to the WTC was $500 million.

The terrorists who bombed the WTC expected to bring down both towers. They failed on that occasion. Notes found in Yousef's apartment in the Philippines proved that plans were already being formed to attack

The twin towers of the World Trade Center defined New York City's skyline. (© Image 100/Royalty-Free/CORBIS)

again. They included using chemicals and poison gas against mass populations. One sinister entry bore a frightening resemblance to what later happened on September 11, 2001. It was a scheme to blow up 11 U.S. commercial planes in one day. Yousef hoped to use a new liquid explosive that could pass metal detectors at airports.

World Trade Center Destruction New York City September 11, 2001

The second and much more deadly terrorist attack on the World Trade Center (WTC) on September 11, 2001, combined as it was with a simultaneous assault on the Pentagon, created an altogether new situation for the U.S. government. The 1993 bombing of the WTC, while deadly, was a single event organized by a small group of terrorists who were in the United States. The events of September 11 were at a new scale of violence: multiple attacks on the heart of the nation's economic and military

Rescue and recovery operations continue at the site of the collapsed World Trade Center. (© Bri Rodriguez/FEMA News Photo)

power. Furthermore, it was organized from abroad by well-known enemies of the United States.

CAUSES

American Airlines Flight 11 left Boston for Los Angeles on the morning of September 11 with 92 passengers and crew aboard. Sometime shortly afterward it was taken over by five passengers who were hijackers. Just before 9:00 A.M. it crashed into the upper floors of the WTC's north tower. Fifteen minutes later a second plane, United Airlines Flight 175, also bound from Boston to Los Angeles, hit the upper part of the south tower. It too had been taken over by hijackers. The planes were flown into the buildings at full speed in what can only be compared to the kamikaze tactics used by Japan in World War II in which young pilots crashed their bomb-laden planes into U.S. ships.

CONSEQUENCES

Flames engulfed the upper floors of both towers within moments, and every branch of New York's fire and rescue organizations sprang into

An aerial view of Ground Zero shows the enormous progress made on cleanup of the site six months after the World Trade Center attack. (© Larry Lerner/FEMA News Photo)

action. It was a chaotic situation, and the city's fire and rescue organizations knew they faced a daunting task. The places where they were needed most were above the 80th floor, and they knew that both electricity and elevators would soon be cut off there. Fortunately, there were only 14,000 people in both towers at the time of the explosions, far fewer than in the 1993 attack. Later in the day there would have been three times that number.

Those inside first experienced a gigantic blast and felt the towers swaying. Sprinklers came on as electricity failed and lights went out. For a time the elevators below the 80th floor continued to operate, and many people were able to get into them. Fires started in various places, many of them triggered by aviation fuel then sustained by the flammable materials in the offices. Thousands of pieces of glass, papers, debris, soot and ash—even clothing and body parts from the airplane passengers—rained down on the streets below. Temperatures reached thousands of degrees in parts of the towers as 10,000 gallons of aviation fuel ignited, levels of heat that even the central steel pillars could not withstand.

For about an hour the main supports of the towers held firm, allowing many people to escape. Fires, sustained by chairs, desks, and other flammables, raced up from the level at which the planes struck to the 20 or more floors above, steadily weakening the main steel supports. At those heights the steel was thinner as the total weight to be supported was much less than lower down. Finally there came a general collapse as the upper floors buckled and sides caved in. Like battering rams in ancient warfare, successive masses of thousands of tons of steel smashed down onto the floors below until they could no longer absorb the pressure. Both towers gave way in a cloud of dust. The noise of hundreds of thousands of tons of steel crashing down could be heard all over southern New York City as people ran from the scene as fast as they could. All public transportation had stopped. Among the most horrific of all the things that occurred was the sight of people jumping to their deaths from the top floors to avoid being incinerated.

The scale of destruction and the reckless indifference to civilian life rightly identified the event as war, a new kind of war, and subsequent actions in Afghanistan and elsewhere were in keeping with that analysis. The first response by the U.S. government was to stop all flights in U.S. airspace in case further attacks might be in process. Incoming planes from other countries were routed to neighboring countries. Canada, because of its proximity to the United States, received most of these flights, and for a time its airports were filled to overflowing. The pilots were not informed of the reason for the changes and were told only where to go. It was the thought that unnecessary panic would be avoided by maintaining silence until the planes were on the ground.

The towers had been designed to withstand an impact from a modern jet plane, but not an impact that involved maximum speed and a maximum amount of fuel. Because the flights that were hijacked were meant to fly to Los Angeles, they were fully loaded with fuel. Modern steel skyscrapers had never previously collapsed because none had ever been subjected to the levels of stress imposed on the WTC. It was feared at first that as many as 6,000 people might have died in the towers. Later it became clear that the count was close to 3,000. Among them were 350 fire fighters who had climbed up into the towers to help. This was 30 times the greatest single loss of life ever previously experienced by the city's fire departments. The fires raged on week after week for more than three months because of the large amount of flammable material available to sustain them.

CLEANUP

Recovery operations, including assessments of neighboring high-rises, were launched immediately. Four high-rises next to the towers collapsed and four partially collapsed. Major structural damage occurred to a dozen others. More than a million tons of debris had to be removed, and at a rate of 10,000 tons a day it took several months just to clear the site. Some individual pieces of steel weighed 25 tons. Excavators with a reach of 100 feet and cranes that could pick up as much as 1,000 tons were needed for the work. All of this debris had to be hauled by barge or truck to a landfill location on Staten Island. Nothing at this scale had ever previously been tackled, and costs for the whole project soared beyond $1 billion.

In 1993, following the bomb attack, it took six hours to get most of the people out of the towers. Those who worked above the 70th floor found it was slow and difficult to make their way down to ground level. Exit stairways were pitch black, and people kept bumping into walls or one another on the way down. Subsequently, batteries were added to every light fixture in stairways for power outages. A public address system was added to enable workers at fire command stations to address evacuating occupants. Fire drills, which previously had often been ignored, were faithfully attended after 1993. Because of these improvements, people were able to evacuate the buildings in 2001 at a faster pace than before. The total time for exiting was cut by a third, from six hours to two hours, and many lives accordingly were saved.

The dangers from toxic materials at the time of the attack were largely ignored because more urgent matters commanded attention. Everyone near the towers as they collapsed was covered with dust from fiberglass, computer screens, asbestos, and a host of products made from chemicals. Spills of mercury, dioxin, and lead were all around. Some initial testing was done after a week, and it showed the levels of toxic chemicals as being below danger standards. Few local residents were satisfied with these results. They continued to wear masks and protective clothing and pressed the Environmental Protection Agency (EPA) to publish precise data on its measurements. Six months after September 11 the EPA had not responded to their requests, and there was persistent concern over what people thought were remnants of carcinogenic materials.

In March 2002, the American Society of Civil Engineers published its report on the fall of the towers. Its members had spent many days investigating the steel supports at the dumpsite, and while they concluded that no changes should be made to building codes as a result of the tragedy, they did make recommendations for the future. They felt that the connectors holding trusses to walls ought to be tested as rigorously as the main walls. This was not done during construction of the WTC towers. They also felt that fire-resistant material could have been used on many internal areas instead of drywall. Their most significant recommendation was that skyscraper height be limited to 60 floors instead of the 110 of the twin towers.

SELECTED READINGS

American Media. *The Day That Changed America*, ed. Jim Lynch. Boca Raton, Fla.: American Media, 2001.

Dwyer, Jim. *Two Seconds under the World: Terror Comes to America*. New York: Crown Publishers, 1994.

Guillemot, Jeanne. *Anthrax: The Investigation of a Deadly Outbreak*. Berkeley: University of California Press, 1999.

Hawley, T. M. *Against the Fires of Hell: The Environmental Disaster of the Gulf War*. New York: Harcourt Brace Jovanovich, 1992.

8

Toxicity—Industrial

Pesticide Leak
Bhopal, India
December 3, 1984

In the 1960s, the developed world introduced the Green Revolution to less-developed countries such as India. It provided new strains of grain and other crops that increased productivity. By coupling these crops with pesticides to reduce losses from insects and other pests, these countries could thus become self-sufficient in food. Union Carbide's Bhopal fertilizer plant was an integral part of India's Green Revolution. It was designed to manufacture Sevin, a pesticide first developed in 1956 and since then the one most widely used throughout the world. Building the plant in India rather than bringing supplies of fertilizer from North America was beneficial to consumers and manufacturer alike. India gained from the capital investment, and the company got low-cost labor, almost exclusive access to the Indian market, and lower operating costs.

Bhopal, a predominantly Muslim city of 800,000, was selected because of its location. It was centrally located on a railway system that spanned the country. A large lake nearby provided a reliable source of water, and enough labor and electrical power was on hand to sustain a major industrial operation. The city tripled in size following the addition of Union Carbide's fertilizer unit in 1969. The first 10 years of operations at the new plant were highly successful, and adequate safety precautions were in

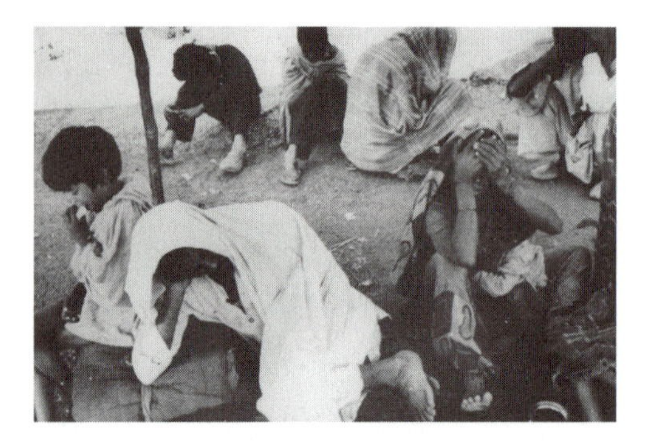

Some of the thousands of people stricken by a poisonous gas leak from a Union Carbide pesticide factory in Bhopal. (AP/Wide World Photos)

place. Indian chemical engineers were taken to the United States for training and returned to their own country to oversee operations and train new staff. By the beginning of the 1980s, however, things had changed. Huge losses had overtaken the company, partly due to lack of demand for pesticides. The Green Revolution was yielding a surplus of food, and there was less need to buy expensive pesticides in order to reduce losses from insects.

As profits slumped, cost-cutting measures appeared. Instead of sending their chemical engineers to the United States for training, men who had taken some university-level science courses were given a four-month crash course locally and then handed major responsibilities within the plant. Because these people were not qualified chemical engineers they could be paid less, thus reducing the budget for staff. For people at this level of responsibility, pay was usually $30 a month, 10 times what an average Indian worker would get. The level of training steadily deteriorated with each new group of workers because none of them had access to new professional developments in the field. Additional workers were frequently needed because many of the best-trained chemical engineers left for better pay and greater security elsewhere.

The men who were hired and trained locally had to work with a very dangerous toxic chemical, methyl isocyanate (MIC), one of three used in the manufacturing process. This chemical could react with several different substances, including water, to produce a lethal gas if great care was not taken during the production process. Unfortunately, few of the workers knew much about it. Until 1979 Union Carbide at Bhopal had imported its MIC as needed from North America. Then in 1980 it began to store this chemical on site for the same reason that lay behind other decisions of that time: to save money. It was cheaper to store it at the plant than bring it in. Two tanks of MIC, each able to hold 15,000 gallons, were added to the fertilizer complex. The decision was controversial because there had been strong opposition in 1969 to installing the fertilizer unit so close to a thickly populated area. The city had laws that would have prevented this kind of development, but political decisions overruled them.

CAUSES

One of the MIC tanks had a leak due to a faulty valve that had been discovered seven days earlier and was not yet repaired. At about 9:30 P.M. on December 3, 1984, a supervisor who knew about the leak in the tank told a worker to thoroughly wash a 20-foot section of pipe used to fill the MIC tanks. The pipe also had a leak, but the supervisor, knowing this, told the worker to clean it. For two hours the worker flushed water through the pipe, onto the floor, and into the drains, all the time leaking water in different directions. By 11:00 P.M. someone noticed that pressure in the MIC tank had risen to five times what it was at 9:30. Water had been entering the tank since that time, and a chemical reaction had been developing rapidly, with temperature and pressure levels rising to dangerous heights.

Close to midnight everything gave way and a dark cloud of 40 tons of gaseous MIC was forced out of the tank into the immediately surrounding community. A hundred gallons of water had entered the tank while the pipe was being cleaned, and the chemical reaction was therefore massive. Furthermore, the tank was already filled to capacity, something that would never have happened if the workers had known the safety rules. A further complicating factor was the operative temperature when the reaction began. The MIC was supposed to be kept at a very low temperature at all times in order to minimize the risk of a reaction, and for that purpose a refrigerator had been installed beside the tanks. It had been switched off for the previous five days for maintenance. When the gas cloud finally burst out from the MIC tank, the supervisor failed to take any action for an additional two hours. By then the gas had spilled far out into the surrounding area (see figure 8.1). Because MIC is a gas that is heavier than air, it stayed close to the ground, invading everyone and everything along a five-mile path.

Figure 8.1
Site of Union Carbide's Bhopal Fertilizer Plant (Paul Giesbrecht)

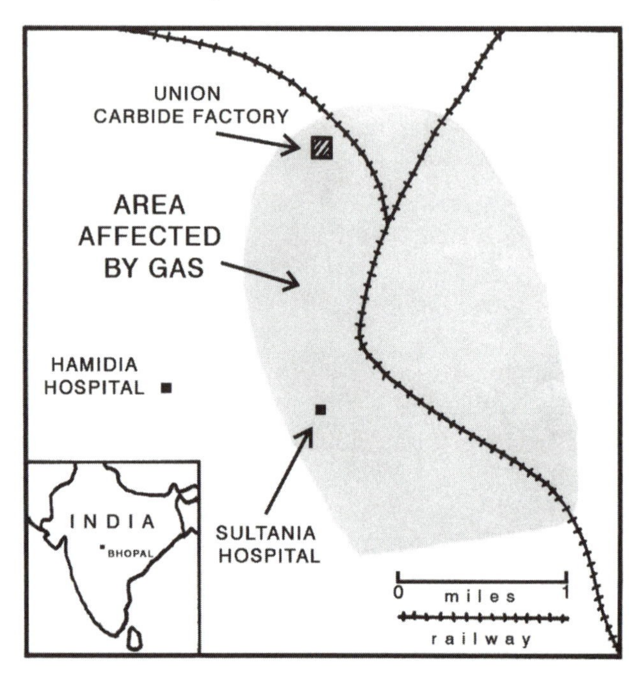

CONSEQUENCES

Throughout Bhopal, thousands of people were killed in their sleep. A substantial number had warning of the accident and were able to escape. The most seriously affected areas were the densely populated buildings close to the plant where the poorest lived. They were the ones who suffered most. Symptoms among the people affected by the poisonous gas took different forms depending on their distance from the factory. They included immediate irritation, chest pain, and breathlessness. Without intervention the problem developed into asthma, pneumonia, and finally cardiac arrest. Almost nothing was known by the population at large about what to do in the event of a chemical leak. In fact, simple protective measures such as placing a wet cloth over the nose and mouth could have saved many lives.

Over the following days and weeks symptoms of poisoning showed up in new ways. Chronic bronchitis and emphysema were found in many who had inhaled the gas. Gastrointestinal and ophthalmic problems were common, and a variety of neurological disorders appeared. Anxiety and depression appeared in different degrees of intensity. Most disturbing of all was the uncertainty about long-term effects. Misdiagnosis was common because local doctors had no experience with these types of illnesses. The total number of deaths was estimated at 4,000 and the number of injured exceeded 50,000. Ten years after the tragedy at least 50,000 people were partially or totally incapacitated as a result of their contact with the MIC gas.

CLEANUP

It became clear in the course of investigations that safety standards and maintenance procedures at the plant were totally inadequate. They had been deteriorating for the three years preceding the tragedy and had caused six minor accidents, all related to the MIC tanks. Gauges measuring pressure and temperature in these storage tanks became so unreliable that workers no longer trusted them and ignored any danger signals they might indicate. A gas scrubber designed to neutralize escaping MIC gas was shut down for maintenance. A flare tower had been installed to burn off gas that might escape from the scrubber, but it too was turned off because a corroded piece of pipe needed to be replaced. Other safety devices and warning systems were also out of service. While there was some doubt about the ability of these safeguards to neutralize the massive amounts of gas that escaped, any one of them could have reduced the death toll.

Fallout from the accident was felt across the chemical industry. Safety audits and new regulatory standards became a primary focus of government and industry. Nongovernmental agencies increased their public awareness campaigns to ensure there would never again be another Bhopal. Concerns about technology transfers, the relations between economic and environmental issues, and the interests of workers all led to intense debate over public policy. In India, the Disaster Management Institute was formed to provide long-term planning to prevent future industrial accidents. The chemical industry responded with the formation of the Center for Chemical Process Safety to develop management strategies for the industry.

Within a year of the gas leak, in response to decisions by Indian courts, Union Carbide agreed to pay a total of $470 million to the victims, but little action was taken for almost a decade because of continuing legal claims. By 1991, six years later, this sum of money had become $700 million as a result of accumulated interest. Finally, early in 1993, the distribution of compensations began. Relatives of the dead received $3,200 for each fatality and more if there were several deaths in one family. People who suffered serious injuries were given $3,000 each, and those with minor injuries substantially less. Union Carbide's stock prices fell precipitously in the years following the accident. The company's reputation suffered badly and it had to sell off many of its holdings in order to survive. By 1992 it was reduced to half of its former size. Early in 2001 it ceased operating and was bought by the Dow Chemical Company for $16 billion.

An interdisciplinary team of specialists appointed by the Bhopal authorities in 1999 investigated the long-term effects of the 1984 tragedy and presented its findings in 2002. Toxic chemicals had migrated more than two miles in all directions from the factory into residential areas and groundwater. From these sources came contamination affecting breast milk, water resources, and vegetables. Dangerous levels of mercury, lead, and nickel were found. Groundwater samples showed the highest concentration of toxic substances. Concerns were expressed that the many survivors from 1984—people still needing medical help—might infect the next generation.

Dioxin Poisoning
Seveso, Italy
July 10, 1976

The Icmesa Chemical Plant at Seveso, Italy, near Milan, was built by the Swiss company Hoffman-LaRoche to produce herbicides and medicines. The chemical processes involved created quite a lot of heat, and this heat was prevented from rising too high by evaporating another chemical alongside it to reduce temperature. Once a batch of the finished product was completed, all activity was brought to a stop by immersing the lab components in a cold-water tank. The whole manufacturing sequence was thus maintained at a very low temperature, an extremely important consideration. If temperatures at any time rose above a certain, critical level, a new and dangerous compound, dioxin, would form. Dioxin, also known as TCDD, is one of the most toxic substances known.

The danger of this happening was overlooked when a manager at the factory decided to change the original design. The alternative he chose was to use water only for cooling at all stages rather than the chemical that was being used for temperature control. The revised system required more time and substantial quantities of water to keep temperatures as low as they were kept using the former method. The new cooling method was not explained to workers at the plant. They were accustomed to seeing

Greenpeace activitists demonstrate at the entrance to the International Conference Center in Geneva, Switzerland, Monday, September 6, 1999, prior to the opening session of the International Chemical Treaty Negotiations. The two children flank a picture of a victim of the Seveso dioxin disaster in Italy. (AP/Wide World Photos)

the very high temperatures come down within 15 minutes using the original method, and they expected similar results with the new approach.

CAUSES

The accident that led to the formation and release of dioxin happened on Saturday morning, July 10, 1976, when workers went off shift oblivious to the fact that the hot chemicals needed additional water to lower the temperature. These workers were unaware of the terrible consequences of overheating, and they thought that it was safe to allow the cooling to take place slowly over the weekend when the factory was closed. In the vents above the reactor where the two chemicals were mixed were rupture-discs that would open under pressure and then close as pressure dropped. In the original design, they were spring-loaded but management had removed the springs so that they stayed open after opening.

As heat and pressure built up in the absence of water, dioxin formed. The rupture-discs opened and a cloud of gas carrying more than two pounds of lethal dioxin spread out into the surrounding area. It was not accompanied by any noise that might have alerted people to its presence. All that was visible was a small cloud rising above the factory. Because residents in the area were accustomed to various odors emanating from the plant, for a time they ignored the greater volume of gases they were experiencing.

CONSEQUENCES

A breeze blew the gas over Seveso and several communities bordering Milan. To avoid the unpleasant gas plume, several people went home and shut their doors only to find that the gas was already inside. In the course of the night they suffered headaches and nausea, and the following morning their children and they had swollen eyes and skin blisters. Their doctors were unable to say what the problem was. For more than a week none of the residents of the affected area knew that the cause of their troubles was dioxin, and even if they had known, there was little knowledge in the local medical community for dealing with it.

It was not long before animals around Seveso were dying by the thousands and leaves on the trees were withering. Doctors and hospitals were swamped with patients who had skin problems. People were being evacuated from the most contaminated area (see figure 8.2), and consumption of all local produce was banned. About one week after the silent explosion, Icmesa officials urged a mass evacuation from the whole contaminated area. The reality was becoming clear. There had been a medical catastrophe and its effects were likely to last a long time. One regional health officer declared that Seveso had experienced its own Hiroshima.

The heaviest blow of all came still later, after the gas cloud was gone. A report from the U.S. Food and Drug Administration made it clear that dioxin, even in very small doses, can damage kidneys, livers, and lungs. It is also extremely dangerous for fetuses, and the fear of having deformed children swept across Italy. Doctors in Seveso warned that if they found deformed fetuses in pregnant women they would recommend abortions. This caused heated debate everywhere because Italy is a Catholic country and the Roman Catholic Church opposes abortion. On August 2, Italy's government revised its strict laws against abortion to permit them for the women who had been affected by the dioxin leak in Seveso.

CLEANUP

To stop the spread of contamination, an army of veterinarians wearing protective suits destroyed all the surviving animals in the affected areas. This was followed by the destruction of cornfields and vegetable gardens, but no one knew how to guarantee a complete cleanup. Reports from other

Figure 8.2
Dioxin-Contaminated Gas Blew over Seveso and toward Milan (Paul Giesbrecht)

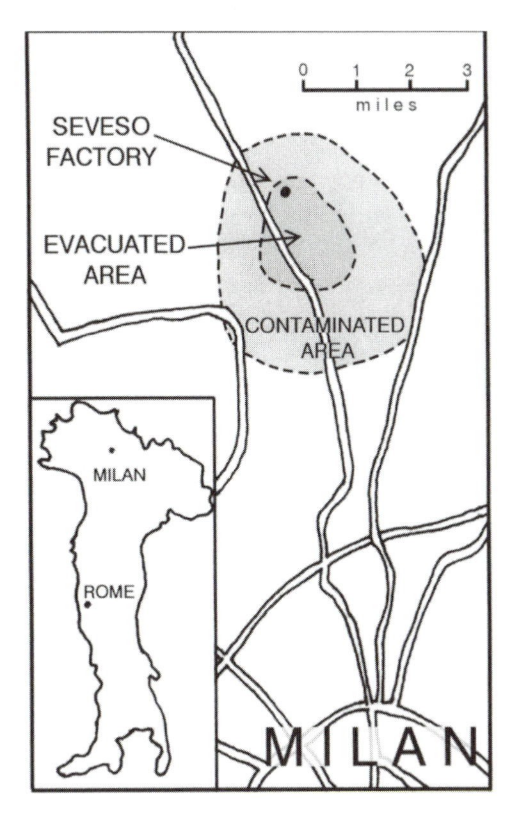

countries were not encouraging. In Britain eight years earlier, a dioxin spill had occurred at a chemical factory. The only solution there was to pull down the whole factory building and bury it deep in an abandoned mineshaft. Vietnamese experts who had to deal with the effects of defoliation from the Vietnam War of the 1960s and 1970s warned that it would be difficult to limit the contamination to the area around Seveso. One of the main components of the defoliant employed in that war was dioxin.

The plant was permanently closed in September because no one believed that the contamination inside it could ever be removed. A large area around it was fenced off with barbed wire. Soldiers wearing protective clothing, rubber boots, and gas masks patrolled the area day and night. Although the cloud of gas was gone within an hour of the accident, the dioxin continued spreading through people traveling, birds flying—

even the tires of cars conveyed the poison to other locations. The Seveso River could carry contamination to the River Po, of which it is a tributary, and so affect the largest farming and industrial regions of Italy.

One valuable outcome of the Seveso disaster was the creation of the European Community's (EC) Seveso Directive, a new system of industrial regulation. Within the EC, each country previously followed its own rules for managing industrial safety. Urgent discussions about a new EC-wide regulatory framework for ensuring the safety of hazardous installations started after the tragedy at Seveso. Neither the residents of Seveso nor the local and regional authorities suspected that the Icmesa plant was a source of risk. They did not even know much about the type of production that occurred there.

The factory had been in existence for 30 years before the disaster, and the only occasional complaints from nearby residents related to unpleasant smells. Of much greater significance were the design changes that compromised the safety of the facility and the population but were not communicated to the authorities responsible for public health and safety. In light of this disastrous accident it was clear that new legislation was needed to improve the safety of industrial sites, to plan for off-site emergencies, and to cope with the broader regional aspects of industrial safety.

The Seveso Directive, adopted by the Council of Ministers of the European Community in June 1982, is the result of those efforts. A central part of the directive is a requirement for public information about major industrial hazards and appropriate safety measures in the event of an accident. It is based on the recognition that industrial workers and the general public need to know about hazards and safety procedures. This is the first time that the principle of "need to know" was embedded in European Community legislation.

Although the Seveso Directive grew out of deficiencies in the existing system of industrial regulation, it is not only intended to provide protection against hazards. It also serves to equalize the burden of regulation on industry. The creation of a single hazardous-industry code leveled the playing field for trade within the European Community by depriving unscrupulous industrial operators of competitive advantages they might gain by exploiting differences among individual countries.

Twenty-five years after the events of 1976 there is a 300-acre park, Seveso Oak Forest Park, where the Icmesa Chemical Plant once stood. It is a popular picnic site. Beneath it lie the poisonous remains of the dioxin spill stored in two enormous concrete tanks. They contain the top 16 inches of soil from all contaminated areas, the bodies of animals that had to be slaughtered, and the factory that caused the tragedy. It was taken apart brick by brick by workers in protective suits and placed below ground in the concrete tanks. Water periodically seeps from the tanks into another container where any dioxin remnants are treated.

Mass checks were made on 10,000 of the people most exposed to the dioxin. In several hundred of them there was a significant reduction in the body's immune response. One extraordinary finding emerged over the seven years that followed the spill. Women who had experienced some contamination, but not sufficient to require an abortion, gave birth to more female babies than males babies, at a ratio of 46 females to 28 males among the whole population of births. This was the first discovery of a molecule that could change the sex ratio.

In 1976 Dr. Paolo Mocarelli was put in charge of a lab to test affected people. He decided to take a blood sample from each of the 30,000 most affected and keep these samples in refrigerated storage in the hope that one day a test would be developed to determine levels of dioxin from a person's blood. That discovery was made 11 years later, and as a result of Mocarelli's foresight Seveso is now a world capital of expertise regarding dioxin's effects on humans. Twenty-five years of patient records coupled with original blood samples are available to researchers.

"Minamata Disease" (Mercury Poisoning)
Minamata, Japan
May 1, 1956

Minamata is a small community on the western shore of Kyushu Island in the south of Japan. Fish was traditionally the main part of the people's diet, supplemented with farm products. The Chisso Corporation's factory, built there in 1907 to produce fertilizer, was converted to a petrochemical plant in 1932, manufacturing materials for the plastics industry. The particular chemical used in its production process contained large quantities of mercury. Chisso was a successful company throughout its history, and because its factory was the only major industry in what is now the city of Minamata, the town grew as the company prospered.

CAUSES

Waste materials from the Chisso factory, including organic mercury compounds, were regularly dumped into the Minamata River and Bay. Little was known in Minamata about the dangers of mercury poisoning, so nothing was done about the health risks that might arise from eating fish

A crowd of demonstrators carrying banners and signs marches in Kumamoto, Japan, in support of court proceedings agains the Nippon Chisso Hiryo Company, which is accused of dumping industrial mercury wastes which cause Minamata disease. (© Michael S. Yamashita/CORBIS)

from the bay. Mercury poisoning was well known in other parts of the world. In nineteenth-century England it caused problems among workers in the hat-making industry. A mercury-based product was in use at that time to stiffen the brims of beaver-skin hats. In their daily work, hatters inhaled the mercury fumes and in time suffered brain damage, in some cases to such an extent that they became insane.

CONSEQUENCES

For about 20 years the waste materials from the Chisso factory did not cause any apparent problems in the health of people. By the mid-1950s, however, health concerns dominated conversations. Unusually large numbers of mentally retarded children were born, and a number of illnesses that were formerly unknown in Minamata appeared. When autopsies were performed on several people who died, a substantial loss of brain cells was detected. A general deterioration was occurring in the brains of many residents, leading to numbness in body extremities, slurring of speech, damage to vision, and even motor-function failure so that people could not walk.

Finally, in spite of denials from Chisso, Dr. Hajime Hosokawa of the Chisso Corporation Hospital declared on May 1, 1956, that the various outbreaks of illnesses came from a common source—mercury contamination of fish that subsequently were eaten by the people. The name *Minamata disease* was given for the first time to the various health problems experienced. The range of debilitating problems continued to mount. Serious brain damage was diagnosed in significant number, including experiences of unconsciousness and the appearance of involuntary movements. These included arm and leg movements as well as uncontrollable shouting. Cats and dogs were affected in ways similar to humans. Birds dropped to the ground from time to time and remained there, unable to move. A general sense of panic was evident throughout the town, made worse by the company's indifference and the unpredictability of the affected people's future.

Despite all the evidence that Hosokawa had assembled in 1956, nothing was done by the Japanese government for 12 years to recognize the validity of his work or to help the disabled. The company continued its production and dumping of wastes into the river; it also continued to maintain that it was in no way responsible for the medical problems. Chisso went further and, by promising financial and other forms of assistance, persuaded some of the affected people to sign documents absolving the company from any liability for their illnesses.

CLEANUP

At last, in 1968, the government of Japan acknowledged that mercury poisoning was the cause of the tragedy and began to consider ways of compensating victims. The Chisso Company ceased production of mercury-based chemicals the same year. Some indication of the level of mercury poisoning taking place can be gauged from the following: The average amount of mercury in fish in the open sea is 3 parts per million but those in Minamata Bay have 50 parts per million. A level of 9 parts per million is considered dangerously high. By 2000 more than 2,000 people were certified as having Minamata disease. Eight thousand more were listed as noncertified patients.

By 1997 Japan's Ministry of Environment finally declared Minamata Bay safe. Over the previous few years it spent almost $400 million in dredging the ocean and river bottoms to remove all mercury-contaminated mud and place it in landfill. Since then the city of Minamata has been sending out messages to developing countries, particularly in Asia, warning them not to repeat Japan's mistake. As part of these messages, government officials from several Asian countries were invited to attend a one-month training course on the environment. In 2001 representatives of nine countries accepted the invitation.

Rhine River Poisoning
Basel, Switzerland
November 1, 1986

The Rhine River runs through the most populated and most industrialized part of Europe, and over most of the twentieth century its pristine waters were so contaminated that aquatic organisms almost completely disappeared. It is a major waterway for Europe's commerce and travel. Its source is in the mountains of Switzerland, and from there it flows past Basel on its 500-mile journey to the North Sea. There are 50 million people living very near the river. Damage reached a peak during World War II as armies crossed and recrossed the river, further degrading its waters and life forms. Then, in 1950, five countries—France, Germany, Luxembourg, the Netherlands, and Switzerland—decided to do something about the problem. They formed the International Commission for the Protection of the Rhine.

For some time the commission had little success. The money needed to clear up the river was unavailable, and there was not enough political pressure to make the problem a high priority. It was not until the 1970s, when the state of the river began to make news all over the world, crippling Europe's tourist industry, that billions of dollars began to come in from the nations bordering the river. Pollution controls on industries and cities were tightened and strictly enforced. Industrial and population growth increased concurrently but, nevertheless, the efforts of the reformers made significant gains and life began to return to the river.

CAUSES

Early on the morning of November 1, 1986, a fire broke out in a warehouse of the Sandoz Chemical Company's factory in Basel, which bordered the Rhine River. It was spotted by a patrol officer who was able to alert the nearest fire department at once. Flames were shooting high into the air as the fire fighters worked to control the blaze. Some of the chemical products in storage continued to fan the flames. It took five hours to bring the fire under control, and by that time the warehouse was totally destroyed. Many thousands of gallons of water were used in the course of this work, and the water, laced with toxic chemicals, poured into the Rhine River. About 30 tons of poisonous chemicals, including hundreds of pounds of mercury, were included in the black tide that moved downstream.

CONSEQUENCES

Within two days the catastrophic effects of the toxic soup were affecting life all the way to the sea. Members of the commission that had done so much to remove pollution from the river were devastated. Sheep that drank from the river became sick and died. Dead fish were everywhere. Water treatment plants in several German villages closed their floodgates and had their fire departments bring in fresh water. Protests, at times violent because the Sandoz Company took more than a week to acknowledge its responsibility for the tragedy, occurred in all the towns and villages bordering the river. At that point Switzerland offered to pay compensation to the people who had been affected.

CLEANUP

Two months after the spill, a group of independent Swiss technical experts published the first assessment of biological conditions on the Rhine. Their report indicated that the microbiology of the river had survived the toxic wastes. The river, particularly where stream flow was high, appeared to have washed out pollution residues. Small, invertebrate water life and plants were alive and they, in the opinion of the experts, would soon provide the basis for a regeneration of fish stocks.

The enormity of the tragedy created a sense of urgency among the signatories to the commission and raised public support to a new height. Scientists predicted it would take at least 20 years for a recovery and stated that it was the worst case ever of chemical contamination in a European river. To capitalize on the new awareness, the commission set a goal for the restoration of the river's habitat—the recovery of a salmon population that would thrive all the way from the sea to Basel. This goal was to be achieved by 2000, and public relations campaigns in support of it were launched all over Europe. The first tangible results came in 1990. Salmon were found to be spawning in tributaries of the Rhine at points 150 miles upstream. By 2000 the goal had been achieved. Fish were once again swimming in large numbers, with hundreds being caught all the way from Rotterdam to Basel.

SELECTED READINGS

George, Timothy S. *Minamata: Pollution and the Struggle for Democracy in Postwar Japan*. Cambridge, Mass.: Harvard University Press, 2001.

Johnson, Ralph Whitney. *Cleaning Up European Waters: Economics, Management, and Policies*. New York: Praeger, 1976.

Kurzman, Dan. *A Killing Wind: Inside Union Carbide and the Bhopal Catastrophe*. New York: McGraw-Hill Book Company, 1987.

Whiteside, Thomas. "Contaminated." *New Yorker*, 4 September 1978.

9

Toxicity—Residential

Basement Contamination
Love Canal, New York
August 2, 1978

In the last few years of the nineteenth century, a developer named William Love set out plans for a new model community next to the Niagara River in New York. He was sure that the location would be popular, and he received strong support from several New York State agencies. A canal was to be constructed around the waterfalls. Hydroelectricity for the community would be generated locally using the river's 284-foot drop from upper to lower levels, and the canal would provide transportation for industrial products. Love's dreams never came to fruition and his canal was never finished.

In 1890 direct current (DC) was the only method available for transmitting electricity. It had the disadvantage of high power losses if the source of electricity was not local. Love's community was planned to be very close to its source of electricity. Early in the twentieth century the principle of alternating current (AC) was discovered, however, and AC quickly replaced DC because it could transmit electricity considerable distances with little loss of power. Factories could now be located at a distance from their sources of power. As a result of this development, Love's project was abandoned, and the canal, which had been partly excavated, remained as a ditch 60 feet wide and 1,300 feet long.

The foundation is all that remained on October 14, 1978, of the Heisner house in Niagara Falls' Love Canal neighborhood. The family moved the house itself to another section of the city. Houses in background are boarded up and abandoned. (AP/Wide World Photos)

The site remained a recreational area for many years and was used as a dump for municipal waste. In 1942 the Hooker Electrochemical Company, which had been producing chemical products for some time, bought the area around the canal from the Niagara Power and Development Corporation and proceeded to add its chemical wastes to the canal. There were few homes in the area at this time, and the existence of a bed of impermeable clay five feet below the surface seemed to the Hooker Company to make it suitable as a dump for chemical wastes. The company felt that the clay barrier would prevent any toxic materials from reaching the water table. Over the period 1947 to 1952, about 25,000 tons of chemical wastes, some in sealed drums, were dumped into the canal. The City of Niagara Falls continued to dump waste there too throughout this period.

In 1954 the Hooker Company sold the land to the Niagara Falls School Board. People had been settling in the area and a school was needed. The sale price was one dollar, but as part of the sale agreement the school board accepted responsibility for any chemical wastes deposited on the site, leaving Hooker free from any future liability. The prospect of a school being built provided an incentive for more people to move to the 70-acre subdivision. The canal was condemned as unusable by the school board, and in 1955 a new school was built a short distance south of it. Both school and residences occupied land bordering the old canal on all sides (see figure 9.1). By the early 1970s there were 800 single-family homes and 250 apartments there.

Figure 9.1
Love Canal in Niagara Falls. Shaded area is public housing (Federal Emergency Management Agency, 1982)

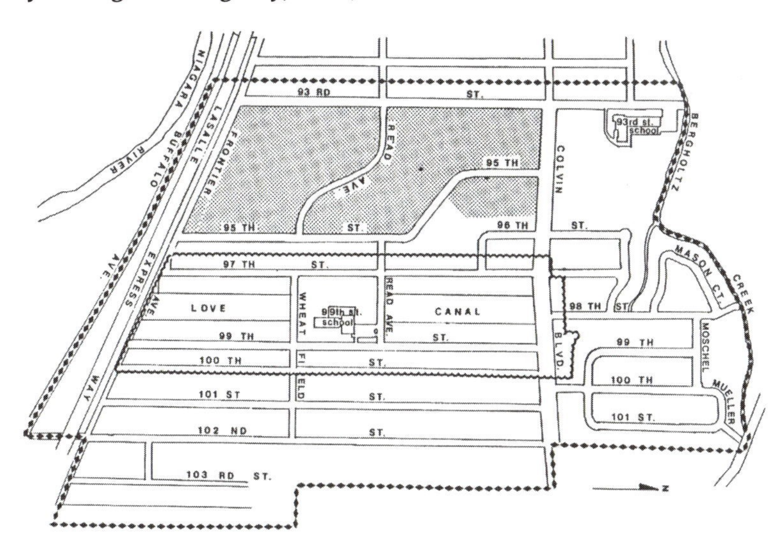

Complaints about toxic wastes soon began to surface. Children who had played around the canal came home with stains on their shoes that would not wash off. A few had burns on their hands from touching chemical wastes. Dogs showed unusual symptoms such as hair loss and skin rashes. A few of them died. Plants and vegetables failed to grow in gardens that were close to the canal. Throughout the areas bordering the canal there was an unpleasant odor. Things got bad in 1975 and 1976 when unusually heavy rain, and snow raised groundwater levels, and 55-gallon drums along with oily solids surfaced from the canal. During these periods of heavy rain basements of the homes closest to the canal were flooded with a watery, black liquid. Sump pumps that were installed to dry out the basements became corroded in a short time and failed to work.

CAUSES

The city of Niagara Falls hired an investigator to examine the grounds and hear the complaints of residents. He soon discovered that the groundwater of the whole residential area had been contaminated with the same black liquid that entered the yards and basements of homes. It was established that drums of chemicals had worked upward to the surface of the canal and leaked. Several toxic chemicals were identified in the groundwater, most of them carcinogenic. A group of residents decided to take action. They formed a homeowners association and collected signatures on a petition demanding that the school be closed down. At the same time, news of the problems at the canal attracted the attention of the state's departments dealing with health and environmental conservation, and they initiated their own investigations.

CONSEQUENCES

The homeowners association pressed its request for the closure of the school with the state after the local school board refused to act. They strengthened their case by hiring Dr. Beverley Paigen, a biologist and specialist in cancer research. She collected medical statistics on the residents and made some alarming discoveries about the homes closest to the canal compared with those farther away: miscarriages by pregnant women occurred three times more frequently, birth defects three times more frequently, and asthma rates were four times greater. Her tests on a random sample of people revealed chromosome damage in one-third of them. This is a condition that is often linked to cancer and can cause birth defects. Paigen's final report listed more than 20 health problems she had found at the site, and in her concluding note she declared that the Love Canal was as much a disaster area as any hurricane or earthquake scene.

Hazardous chemicals were also found by the New York State Department of Health, which declared the Love Canal area a threat to human health. As a result, the canal area was fenced off, the school closed, and arrangements made for pregnant women and young children to be evacuated. When traces of dioxin were found in the soil, there was extreme agitation among all the residents. Dioxin is one of the world's most carcinogenic chemicals. They feared both for their health and the risk that their property values were about to plummet. The community came together and demanded action to deal with the crisis. They finally secured it from the state's health commissioner, Dennis Whalen, in Albany, who on August 2, 1978, declared a health emergency.

Whalen called Love Canal a great and imminent peril to the general public and urged residents to avoid using their basements and not to eat food grown in their gardens. The earlier actions—fencing off the landfill site and evacuating children and pregnant women—were immediately extended. About 240 homes bordering the canal were purchased by the state and the people living in them permanently relocated. At the same time remedial work to contain the wastes from the canal was launched. All of this still left large numbers of other homes close to the canal occupied. Pressure increased from residents for action to protect all who were endangered.

Within three months additional quantities of dioxin were found at some distance west of the canal, the effects of the remedial measures being undertaken. Instead of containing the wastes and limiting their influence, the excavation work became accessible to runoff, and large quantities of pollutants flowed into nearby sewers. More homes were purchased by the state and their occupants relocated. About this time intense political debate surfaced involving local and state authorities and federal agencies, all in relation to how much of the area could be retained for residence. The state concluded that the residential perimeter of the canal was all that need be evacuated. Residents did not agree. They launched legal action and secured the right of temporary relocation for more people.

Meanwhile the quality of life for remaining residents deteriorated. Homes that had been abandoned and were now separated from the rest of the area by a chain-link fence became targets for vandals and thieves. Burglaries and fires were common. By late 1977 one member of Congress and the federal Environmental Protection Agency (EPA) became involved, and the latter began to examine the basements of the homes closest to the canal. Feelings were running high among the residents because of the many delays. Many protested, and on one occasion federal EPA officials were held captive for several hours. Finally, in May 1980, almost two years after the New York State declared a health emergency on August 2, 1978, President Jimmy Carter declared a federal emergency in the Love Canal area, thus clearing the way for the relocation of the remaining families.

CLEANUP

Following the declaration of a federal emergency, the Love Canal Area Revitalization Agency was formed with the support of federal money. Its task was the restoration of the canal area to livable status. All toxic wastes and abandoned homes were to be removed. Decontamination work began in 1981, but with frequent impediments to its progress. There were lawsuits as different individuals and groups sought compensation and the various political agencies and levels of government squabbled over their rights and recognition. By May 1982 the federal EPA was convinced that the area was habitable, and plans were gradually advanced for resettlement. One lawsuit by the federal Justice Department forced the Hooker Company, by then known as the Occidental Petroleum Company, to pay part of the costs of restoration.

The EPA's final action on the Love Canal came on October 28, 1987, in the form of a $30 million federal grant. This money was to be used for on-site thermal destruction to clean up the dioxin-contaminated creek and sewer sediments and to cover the costs of disposing of the residuals from the thermal-destruction process on site. A portable thermal-destruction unit was installed at Love Canal for this cleanup work. On the twentieth anniversary of President Carter's emergency declaration, the scene around the old canal is very different. The canal is safely buried, fenced off from the rest of the subdivision, and declared permanently out of bounds. Dozens of homes are also buried there.

A public corporation assumed ownership of the abandoned properties, fixed them up, and arranged resales. Of the 239 homes it renovated, 232 were sold by 1998. The community has a new name—Black Creek Village. The new residents feel safe in their homes, convinced that every part of their neighborhood was repeatedly tested and retested before they moved in. A recent declaration by the New York State Department of Health assured the people of Black Creek Village that no evidence exists that would link their community in any special way to illness.

Food Poisoning
Iraq
September 1971

The area known today as Iraq was, in ancient times, one of the first places in the world to grow wheat, but because it is also a region of low rainfall, it frequently experienced seasons of drought that ruined the wheat crops. The year 1971 was one of those very dry years, and Iraq decided to change to a new strain of wheat that would be more resistant to climatic shifts. Mexico, also a country of low rainfall, had developed wheat of this kind, and Iraq ordered a large quantity of it in the late summer of 1971. A shipment of more than 70,000 tons was delivered to the port of Basra in southern Iraq, and from there it was distributed throughout the country.

CAUSES

The grain was treated with a fungicide, an organic mercury compound called methylmercury, in order to protect against rot and insects. This treatment is harmless if the grain is used for seed but poisonous if it is eaten. Warning statements were marked on the sacks in Spanish. The grain was sprayed with a red dye as an additional warning but this, like the Spanish words, carried no significance for the Iraqi workers. Everyone who handled the grain shipments was warned of the dangers of eating the grain, and this warning was also relayed to all districts of the country that received grain deliveries. The quantities allocated to various regions corresponded to the amounts they had used for seed in previous years.

Despite the warnings printed on the sacks of grain, the dye on the grain itself, and the announcements from the Iraqi government, people ate some of the grain. They did not eat a lot because they had to retain enough to use as seed in the following year. Some grain was consumed after being baked into bread, and some was fed to animals. A few people who understood the warnings thought they could make the grain edible by washing off the red dye. Consuming even very small quantities of the mercury can do serious health damage, however.

The seed that was fed to animals carried mercury back to the owners when these animals were killed for food, and thus the level of mercury that had entered their bodies from the bread was further increased. The compounding of the problem through the food chain did not end with the farm animals. When the Iraqi authorities were finally able to convince their people that the whole shipment was poisonous, grain not intended for use as seed was thrown into the Tigris River. Once again the food chain was affected as fish ate the grain thrown into the river. When those fish were caught and eaten, more mercury was ingested.

Symptoms of trouble were not evident for some weeks, but when they did appear they were catastrophic, the worst ever recorded for this type of poisoning in terms of the number of people killed and injured.

CONSEQUENCES

The consumption of bread made from the grain became the main source of health problems throughout the country. All the recorded cases of poisoning occurred in rural areas where bread was made at home. In the bigger cities, where bread is prepared commercially from government-inspected flour, there were no cases of contamination.

The first signs of poisoning were numbness in fingers and toes and other extremities. This was followed by an unsteady gait and—where the quantities ingested were substantial—the loss of coordination to the point that people were unable to walk. Eyes were frequently affected, with difficulties ranging from blurred vision to blindness. Slurred speech and hearing loss were present in many cases. In all of these instances it was evident that brain damage had occurred. Fatalities were the result of failure of the central nervous system. There was little evidence of damage to cardiovascular or digestive systems. Those who were severely poisoned died despite the medical treatment they received.

CLEANUP

The epidemic was so great that the government appealed for medical help from European countries. At that early stage, no one knew exactly what had happened. When medical teams arrived they quickly diagnosed the cause of the epidemic. The situation was difficult to monitor as grain had been shipped to several places inland and there was concern that people might panic. Radio blackouts were enforced to keep the information from spreading in order to allow the government time to take control of the situation.

Within three months of the initial outbreak the number of cases peaked, with hundreds arriving in hospitals daily. Males and females of all ages were affected, with the largest number of cases under the age of nine. There were equal numbers of males and females. Death rates were highest among the elderly and the very young. In order to minimize the destructive effects of the poison, a type of resin was given to patients orally to hasten the elimination of mercury. The longer the poison stayed in the body, the worse the results.

Mercury from treated grain can enter the human body orally, by inhalation, or just by skin contact. The primary mode of reception, oral, was via contaminated bread, meat, and other animal products obtained from livestock that had consumed treated grain, vegetable products stored in sacks that had contained the treated grain, and game birds and fish that had eaten the treated wheat.

In summary, wheat was purchased by Iraq for use as seed but many things went wrong. Careful supervision of deliveries and more intensive efforts to alert people to the dangers of consumption could have prevented the disaster. The official casualty figures listed 6,500 as admitted to hospitals with severe problems. The official number of deaths was 500, but hospital authorities put the number much higher.

Smog Pollution London, England December 5–8, 1952

Air pollution is a major global problem. It makes health problems worse and produces problems where formerly there were none. In various countries of the world at different times it has resulted in disastrous consequences to forests, soils, air, acid-sensitive aquatic organisms, and the surfaces of steel and concrete buildings. Donora, Pennsylvania, in 1948 and Los Angeles, California, over many years are examples of the most severe cases. At present, in the capitals of the world's two most populous countries, India and China, there are frequent experiences of damaging air pollution. Smog pollution, which refers to fog plus smoke, is a long-standing side effect of industrialization. The severity of the problem and the public awareness of it, however, are relatively new. Worldwide, as industry expanded rapidly after World War II, smog pollution followed.

London's history of fogs is well known. Novelists often refer to it, and many of them conjure up a dark, damp, foggy atmosphere to enrich their stories. The London term most commonly used for this condition is *pea-souper*. The tons of sulfur dioxide and other pollutants that come from coal-burning fires and furnaces are the reasons for it. Great Britain was the first country to develop an industrialized economy based on coal. By the middle of the twentieth century, its factories were releasing 9 million tons of sulfuric acid into the air annually. As this acid descended, it corroded metals, decayed stone buildings, killed nutrients in soils, and rotted clothing. London's deadly experience of a serious outbreak of smog in 1952, while rare even in England, served to highlight the global challenge.

November 1952 was colder than average in London, with temperatures staying near freezing and heavy snowfalls occurring. To keep warm, residents of London were burning large quantities of coal in their homes. At that time it was quite common for individual homes to burn several fires, one in each room. Smoke poured out of the chimneys over these coal-fired grates.

CAUSES

A large high-pressure center settled over London on Thursday, December 4, a mass of cold air with a temperature derived from the area over which it passed before reaching London. It happened to be warmer than the ground below, so a temperature inversion followed; that is, the air above the ground was warmer than the surface below it, so all of the emissions from homes and factories were trapped near the ground.

The next day a typical pea-souper began to form because the high-pressure center remained stagnant and its clear air intensified the radiation escaping from the ground, thus further cooling the ground and strengthening the temperature inversion. Air pollutants accumulated at an unprecedented rate, creating the worst fog in living memory.

CONSEQUENCES

The air became a blinding, suffocating cloud of gas, clogging breathing passages and stinging the eyes. The fog was so thick that people standing were unable to see their feet. At Heathrow, London's biggest airport, visibility dropped to 30 feet and all planes had to be diverted to other places. Over the period from December 5 to December 8 the concentration of particles in the air increased tenfold above normal.

It is hard to imagine a metropolitan area of 10 million people coming to a complete standstill, but that is exactly what happened. Every aspect of city life was crippled and, perhaps for the first time, most of its residents discovered how interconnected and interdependent all the parts really were. All modes of transportation except the underground railway system stopped. Even the subway trains had trouble when they emerged above ground along parts of their route.

In streets everywhere, visibility dropped to a point where nothing could be seen beyond a couple of feet. Drivers were confronted with barricades of abandoned cars. One ambulance attendant walked 25 miles holding an open-flame torch to guide the ambulance driver. All flights arriving at Heathrow were diverted to Bournemouth, a seaside town 150 miles southwest of London. Ships unloading food and other essentials had to stop work because of fears that people and goods might fall into the water and no one would be aware of the accident.

Polluted air poured through windows and under doorways. The city became a place of lost and troubled people. Fleets of ambulances were called in from surrounding communities to help with the masses of people of all ages who just could not cope with the attacks on their lungs. They were of little use because visibility made it almost impossible for the ambulances to get to where they were needed. Doctors could not get to hospitals or to patients at home, so they resorted to diagnosing and treating problems by telephone. Fires broke out, but nothing could be done. Fire fighters from one station were unable to get to a burning building 400 yards away. The fire destroyed the building.

Pregnant women were unable to reach hospitals and had to deliver babies in all kinds of places. Electric power was cut off for large sections of the city when staff failed to turn up for duty at control stations. At the main studios of the British Broadcasting Corporation, both radio and television programs were cut short. At a cattle auction animals were dying as they inhaled the poisonous fumes. Some were saved by an enterprising worker who covered their noses with cloths moistened with whiskey.

CLEANUP

Fortunately the city's paralysis lasted no more than four days, yet in that time many lives were lost and considerable damage done to the physical environment. The week ending December 6 there were 2,062 deaths in London, an average figure. The following week, ending December 13, there were 4,703 deaths. The rate continued to be abnormally high for the next two weeks. Altogether 4,000 people, most of them over the age of 40, lost their lives. Another 4,000 died later as a result of breathing the smog. Many thousands more were seriously ill.

As a result of the disaster, new laws were passed. The City of London's Clean Air Act of 1954 was the first, and its demands were augmented in 1956 and 1968 with additional legislation that strengthened the enforcement powers of local and national authorities. Emissions of smoke from coal fires were banned, and both homes and factories were required to convert to smokeless fuels within specified time frames. Today, London's pea-soupers belong to history. Central heating of homes and offices has taken the place of individual room fires, so coal is no longer the principal fuel in use. Where it is still employed in electricity-generating stations, scrubbers are installed to minimize the amounts of pollutants released into the atmosphere.

The effects of acid rain, the sulfuric acid component of smog that does extensive damage to the physical environment, cannot be assessed on the basis of its activity over four days. Nevertheless, the acid's power over time can readily be seen in the older buildings of London. The damage it inflicts is great. In 1990, restoration of the outer walls of Westminster Abbey cost $20 million. Saint Paul's Cathedral is due for similar restorative work. Its outer surfaces are deteriorating at an accelerating rate. All the older buildings of northwestern Europe are similarly threatened by London's smog because wind flows do not respect national borders.

SELECTED READINGS

Ackroyd, Peter. *London: The Biography*. London: Vintage, 2001.

Bakir, F., Damluji, S., Amin-Zaki, L., Clarkson, T., Smith, J., and Doerty, R. "Methylmercury Poisoning in Iraq." *Science* 181 (20 July 1973): 230–239.

Brown, Michael. *Laying Waste: The Poisoning of America by Toxic Chemicals*. New York: Pantheon Books, 1980.

Fowlkes, Martha A., and Patricia Y. Miller. *Love Canal: The Social Construction of Disaster*. Washington, D.C.: Federal Emergency Management Agency, 1982.

Index

About the Author

ANGUS M. GUNN is Professor Emeritus, University of British Columbia, and author of numerous books, journal articles, and media productions on education, geography, and environmental science. His publications include *Patterns in World Geography*, *Man's Physical Environment*, *Habitat: Human Settlements in an Urban Age*, *Heartland and Hinterland: A Regional Geography of Canada*, and *The Impact of Geology on the United States*.

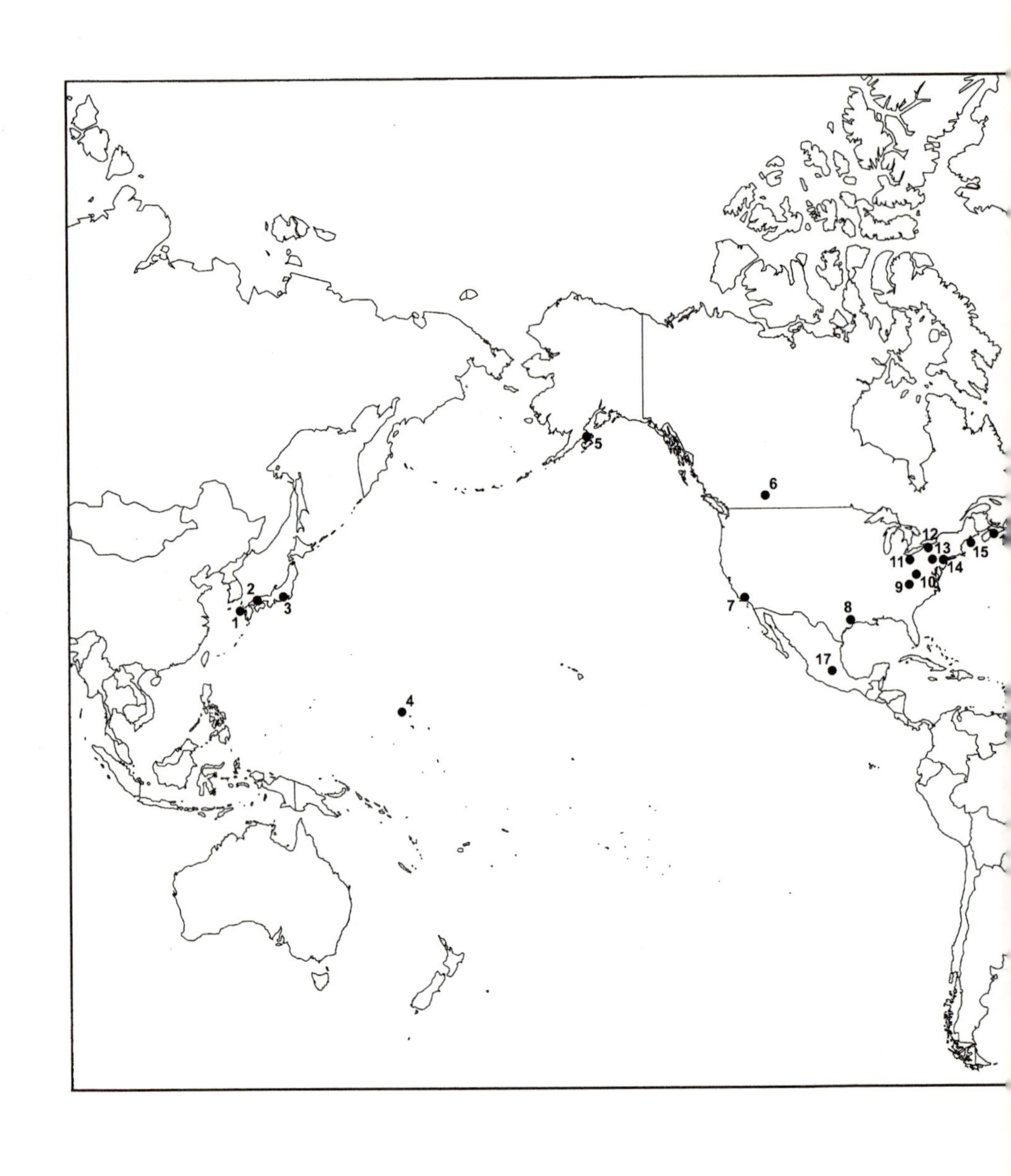
1
2
3
4
5
6
7
8
9
10
11
12
13
14
15
17